William Whiston · Astronomical Principles of Religion

Anglistica & Americana

A Series of Reprints Selected by
Bernhard Fabian, Edgar Mertner,
Karl Schneider and
Marvin Spevack

109

1983
Georg Olms Verlag
Hildesheim · Zürich · New York

William Whiston

Astronomical Principles of Religion, Natural and Reveal'd

(1717)

Introduction by
James E. Force

1983
Georg Olms Verlag
Hildesheim · Zürich · New York

o̲

Note

The present facsimile is reproduced from a copy
in the possession of the British Library,
Reference Division. Shelfmark: 699.g.7.

B.F.

Reprografischer Nachdruck der Ausgabe London 1717
Printed in Germany
Herstellung: Strauss & Cramer GmbH, 6945 Hirschberg 2
ISBN 3 487 07327 7

Linking History and Rational Science in the Enlightenment: William Whiston's <u>Astronomical Principles of Religion, Natural and Reveal'd</u>

Philosophers and historians are notoriously prone to distort the thought of historical individuals by selectively emphasizing only what they want or need from the whole body of his work. This tendency is exacerbated when a person's thought is especially wide-ranging. Many thinkers have suffered anachronistic misinterpretations at the hands of their critics but, until recently, few have fared worse than Sir Isaac Newton and his immediate disciples at Cambridge. Because of the editor's natural tendency to "edit out" and because of the kaleidoscopic pattern of Newton's influence it *is* difficult to delineate all the varied aspects of "Newtonianism." And yet to get an accurate picture of the man, his times, and the whole program of Newtonianism, we must put back in the excised parts. To trace Newton's influence wherever it is found is precisely the "challenge that faces historians of eighteenth century thought".[1]

An examination of the thought of Newton's erstwhile protégé, the controversial mathematician and natural philosopher, William Whiston, presents one fruitful approach to meeting this challenge. Whiston is particularly useful in revealing essential elements of the Newtonian system about which his non-combative mentor is reticent. Frank E. Manuel correctly states that "where Newton was covert Whiston shrieked in the market place".[2]

---

[1] P. M. Heimann, "Newtonian Natural Philosophy and the Scientific Revolution", *History of Science* II (1973), p. 1.

[2] Frank E. Manuel, *Isaac Newton, Historian* (Cambridge, Mass., 1963), p. 143. Manuel has been most instrumental in revising our views about Newton's thought. Manuel's method has been to assign to Newton's ideas the significance accorded to them by Newton him-

Above all, Whiston shows perfectly
how integral to the Newtonian program
is the task of relating a specific historical content to the abstraction of
the scientific theism of the design
argument of natural religion. Newton's
principal concern with theology has
generally been reckoned to be solely
the scientific theism of his version
of the design argument in the "General Scholium" of the second edition
of the *Principia* (1713). But Newton

---

self, especially as this is revealed
in Newtons's vast MSS on Biblical
chronology. P. M. Rattansi's and
J. E. McGuire's article "Newton and
the 'Pipes of Pan'," *Notes and Records of the Royal Society of London* 21, no. 2 (Dec., 1966), pp.
108-43, is an example of the fruit
this procedure bears. More recently,
M. C. Jacob has cast much light on
the social and political implications
of Newton's thought within the cultural milieu of the early eighteenth
century in her excellent volume
The Newtonians and the English Revolution, 1689-1720 (Ithaca, N.Y.,
1976).

is vitally interested in grounding his
religious views not merely in scientific reason, but in history as well.
Whiston, in his *Astronomical Principles
of Religion, Natural and Reveal'd*,
shows how Newtonianism properly includes a complete analysis of the relation between scientific reason and
history. Whiston becomes a biblical
exegete in order to connect the rational natural religion of the design
argument with the revealed religion
of scripture. Whiston labors to save
the historicity of the Bible and to
make that history harmonious with
the Newtonian system. While pursuing
this synthesis between reason and
history, i.e., between "Religion, Natural and Reveal'd", Whiston also reveals the basis for the Newtonian fascination with the "prisca sapientia"
tradition. In analyzing how the two
"Divine Volumes" of scripture, considered as historical testimony,
and science mutually condition one

another the added historical testimony of these "ancient theologians" confirms that the natural religion evidenced by the whole "frame of nature" and the revealed religion of the Bible are in a harmonious synthesis. Finally, Whiston illustrates the Newtonian preoccupation with the coming of the millennium for those who are saved and the torments of those who are not in a uniquely Newtonian vision of hell.

But who is this "downright, upright" William Whiston who so assertively promotes these important but easily ignored aspects of Newtonianism? Who is he, that is, besides the probable model for the Reverend Dr. Primrose in Oliver Goldsmith's *The Vicar of Wakefield*?[3]

---

[3] The similarity between Whiston and Primrose is noted in Leslie Stephen's article in the *Dictionary of National Biography,* s.v. "Whiston, William". Stephen states that Whiston "strongly resembles the Vicar of Wakefield, who adopted his principles of monog-

I. Whiston's Career

Born the son of a parish rector in Leicestershire in 1667 Whiston began his schooling in 1684 at Tamworth School. He entered Clare College, Cambridge, in 1686 and immediately excelled in mathe-

---

amy. His condemnation of Hoadly upon that and other grounds is in the spirit of Dr. Primrose. (*Memoirs,* p.209) It is not improbable that Whiston was more or less in Goldsmith's mind when he wrote his masterpiece." The integrity of the vicar and Whiston and their dogmatic determination to hold fast to the same principles (on the basis of an historical interpretation of "primitive Christianity") in the face of great adversity makes Stephen's conclusion quite likely. Goldsmith is well acquainted with Whiston's many controversies. See Robert H. Hopkins, *The True Genius of Oliver Goldsmith* (Baltimore, 1969), pp. 174-5. Goldsmith also knows Whiston's cosmological work and accurately summarizes Whiston's theory of the formation of the earth. Goldsmith states that it is "most applauded". See Oliver Goldsmith, *An History of the Earth and Animated Nature* (New York, 1825), p. 9.

matics. In 1691 he was elected to a fellowship. He received his M. A. in 1693 and was ordained a priest in the Church of England.

Following his ordination he returned to Cambridge to continue his studies in mathematics and to learn the "*Cartesian philosophy*", but, as Whiston remarks in his memoirs,

> ...it was not long before
> I, with immense pains, but
> no assistance, set myself,
> with the utmost zeal, to
> the study of Sir *Isaac Newton's*
> wonderful discoveries in
> his *Philosophia Naturalis
> Principia Mathematica,* one
> or two which lectures I
> had heard him read in the
> publick schools, though
> I understood them not at
> all at the time. [4]

Whiston's labors gained him a solid reputation as a mathematician and he

---

[4] Whiston, *Memoirs of the Life and Writings of Mr. William Whiston containing Memoirs of Several of his Friends also.* The second edition corrected. (London, 1753), p. 32.

acquired a number of students as a tutor, including the nephew of Archbishop Tillotson, but in 1694 he retired from teaching due to ill health and became the chaplain to Bishop John Moore of Norwich. During his chaplaincy Whiston published his first book, *A New Theory of the Earth, From its Original to the Consummation of all Things* (1696). Whiston showed this work to Richard Bentley and Christopher Wren but it was "chiefly laid before Sir *Isaac Newton* himself, on whose principles it depended and who well approved of it".[5]

In 1698 Whiston was appointed vicar of a parish in Suffolk but held this post for only a short time until Newton, impressed by Whiston's book, asked Whiston to return to Cambridge and to act as his assistant. When Newton left the university in 1703 to become Master of the Mint in London he secured for Whiston his own

---

[5] *Ibid.*, p. 38.

vacated chair; Whiston thus succeeded Newton as Lucasian Professor of Mathematics.

Newton was not the only intellectual impressed by Whiston's writing. John Locke wrote a letter to his friend, William Molyneux, in which he described the opinion of the "ingenious" regarding Whiston:

> He is one of those sort
> of writers that I always
> fancy should be most
> esteem'd and encourag'd.
> I am always for the
> builders who bring some
> addition to our knowledge,
> or, at least some new
> thing to our thoughts. 6)

---

6) Locke to Molyneux, 22 Feb. 1696/7, *Some Familiar Letters Between Mr. Locke, and Several of his Friends* (London, 1708), pp. 176-7. The full passage regarding the "opinion of the ingenious" concerning Whiston's *New Theory* reads: "I have not heard any one of my acquaintance speak of it, but with great commendation, as I think it deserves. And truly, I think he is more to be admired, that he has lay'd down an hypothesis, whereby he has

From the period of his summons by Newton in 1699 until his banishment from the university in 1710, Whiston was engaged in a fruitful academic career. He published in 1703 a new edition of Euclid and Archimedes "with the addition of practical Corollaries ... for use of young students in the university".[7]

In addition to his mathematical and "geological" works Whiston also did extensive research in theology and, especially, Biblical chronology. His research in this field led him to doubt the doctrine of the Trinity on historical grounds and the publication of

---

explain'd so many wonderful, and before, unexplainable things in the great changes of this globe, than that some of them should not go easily down with some men, when the whole was entirely new to all." I am greatly indebted to my colleague, Professor Henry Schankula who is currently editing the Locke Journals for the Oxford University Press, for this source of Locke's remark.

[7] Whiston, *Memoirs*, pp. 114-5.

his heterodox anti-trinitarian views, and his stubborn refusal to recant, led eventually to his expulsion from Cambridge in 1710.

This event forms the turning point in Whiston's career. Up to 1710 he had pursued a successful academic career. After 1710, because of an act of conscience at the age of forty-three with four children to support, he was forced to move down to London and to scramble madly for a living by pursuing a variety of lecturing and publishing schemes. He eventually published a translation of the works of the Jewish historian, Josephus, which remained in print continuously until this century. He was also actively engaged in a variety of attempts to discover an accurate method to determine the longitude of ships at sea. In addition, encouraged by Joseph Addison, Henry Newman, and Richard Steele, Whiston delivered a series of popular lectures on the Newtonian mathematics and system of celestian dynamics which received wide acclaim.

Whiston was introduced to Steele by Henry Newman, Secretary of the Society for Promoting Christian Knowledge. In a private letter from Newman to Steele, Newman recommends Whiston to Steele's patronage and discusses the scheme of Whiston's proposed coffee house lectures. Newman writes to Steele:

> I only beg leave to suggest one thing to you when he does lecture ... and that is that you will be pleas'd to conjure him Silence upon all Topicks foreign to the Mathematicks in his conversation or Lectures at the Coffeehouse. He has an Itch to be venting his Notions about Baptism & the Arrian Doctrine but your authority can restrain him at least while he is under your Guardianship. 8)

---

8) Newman's letter to Steele, dated 10 August 1713, is cited in Calhoun Winton, *Captain Steele: The Early Career of Richard Steele* (Baltimore, 1964), pp. 157-8n.

The effect of Whiston's coffee house lectures on the poetic imagination of one regular attendee in his audience, was profound. In a letter which appeared in the *Guardian* for Sept. 24, 1713, Alexander Pope tells of what he had learned of the importance of Newtonian science as the foundation of an inference from the order and regularity of the Newtonian system to the existence of a providential deity. Pope writes that

> it is a sort of Impiety
> to have no Attention to
> the Course of Nature, and
> the Revolution of the
> Heavenly Bodies. To be
> regardless of those Phae-
> nomena that are placed
> within our View, on purpose
> to entertain our Faculties,
> and display the Wisdom
> and Power of their Creator,
> is an Affront to Provi-
> dence of the same kind. 9)

Whiston, then, was the first Newtonian to give public lectures in which

---

9) Marjorie Nicolson and G. S. Rousseau,

he popularized the Newtonian "frame of nature". It was from Whiston that Pope learned the rudiments of Newtonian natural philosophy and, apparently, its use in the design argument.

In 1717, four years after the coffee house lectures, Whiston published the most important statement of the Newtonian version of the design argument to appear in the first half of the eighteenth century, his *Astronomical Principles of Religion, Natural and Reveal'd*.

Throughout the rest of his long life Whiston managed to make ends meet by his writing and through bequests and gifts from such friends as Sir Joseph Jekyl and Queen Caroline. In addition, his continued public lectures in London and towns such as Bristol, Bath, and Tunbridge Wells enabled

---

"This Long Disease, My Life": *Alexander Pope and the Sciences* (Princeton, 1968), pp. 137-56. The citation from Pope's letter to the *Guardian* is on p. 149.

Whiston to go on with his various studies. Between 1714 and 1740 Whiston proposed three basic schemes for the determination of the longitude but after the publication of the *Astronomical Principles* he turned increasingly to problems of Biblical interpretation.

His *Memoirs* (which read like a trial record rather than an autobiography) show that Whiston loved a fight. As noted above, it was his flagrant infidelity to established Church doctrine regarding the trinity which precipitated his ouster from Cambridge. Even prior to his dismissal he had quarreled with Newton over the proper interpretation of the prophecy of the seventy weeks in the book of *Daniel*. Following his departure from the university, Whiston and Newton became so estranged about the subject of Biblical chronology and prophecy that in 1720 Newton blocked Whiston's admission to the Royal Society. But however violently

they disagreed about the interpretation
of particular Biblical prophecies and
however much the reticent Newton abhored
Whiston's enthusiasm for public debate
(Newton silently shared Whiston's anti-
trinitarian views), the two men ultimately agreed that the book of nature
and the book of revelation were written
by God, and contained one indivisible
divine truth. They were completely un-
troubled by the idea of a warfare
between science and religion.

Of far more concern to Whiston was
his controversies with the deists who
sought to set science in opposition to
religion or, at least, to separate the
two endeavors. Anthony Collins' deistic
*A Discourse on the Grounds and Reasons
of the Christian Religion* (1724) was
directed against Whiston's literal interpretation of Biblical prophecy as
outlined in Whiston's 1708 Boyle Lectures, *The Accomplishment of Scripture
Prophecy*. This topic had originally
been suggested to Whiston by Newton.

Collins' attack provoked a heated, drawn-out response from Whiston in which he argued that the literal, historical fulfillment of Biblical prophecies provided certain confirmation of the truth of Christian revelation just as Newtonian science helped confirm the truth of revelation through the Newtonian design argument.

Beginning in 1726 Whiston began to lecture in public concerning the historical events which would, according to Scripture, precede the millennium. He outlined which of these prophesied events had already been fulfilled and when the others would come to pass. At a lecture in 1746 in Tunbridge Wells, he announced that according to his latest reckonings the second coming of Christ was just twenty years away. Fortunately, the pious controversialist did not live to be disappointed and died "full of years and good works" in 1752.

II. Whiston's Astronomical Principles of Religion, Natural and Reveal'd

Whiston's extraordinary book is dedicated "To the illustrious Sir Isaac Newton, President, and to the Rest of the Council and Members of the Royal Society". It is a textbook example of the way in which a Newtonian scientist utilizes the Newtonian empirical-mathematical description of the "frame of nature" as the basis for an analogical argument for the existence and providential nature of God. And it bears out Manuel's contention that Whiston's work casts new light on the "kaleidoscopic pattern" of Newton's influence.

a. Newton, Whiston, and the Design Deduction

"Natural theology" (or "physico-theology" or "astro-theology"[10] is the most well-known aspect of Newton's theology. Newton repeatedly asserts the design argument in letters and published works. His first words on the subject date from a series of four letters to Richard Bentley in 1692 while Bentley was preparing to deliver the inaugural sermons in the newly created Boyle Lecture series. Newton writes to Bentley a lengthy account of the nature of his celestial system and indicates that the idea of proving a "beliefe of a

---

[10] Another Boyle lecturer who utilized the platform provided by the Boyle Lecture series to promote the design argument was William Derham who used these terms as titles for two very important works in this field. In 1713 he published his *Physico-Theology, or a Demonstration of the Being and Attributes of God from his Works of Creation*. This work was the collection of essays comprising the Boyle Lectures for 1711 and 1712. In 1715 Derham published *Astro-theology, or a Demonstration of the Being and Attributes of God from a Survey of the Heavens*.

Deity" on the basis of the system in the *Principia* had been present in his mind at the time of its writing in 1687.[11]

The last three of Bentley's sermons (which really comprise one extended essay delivered in three parts) is separately entitled "A Confutation of Atheism from the Origin of the World".[12] Bentley closely follows Newton's letters in his presentation of the design ar-

---

[11] Isaac Newton, *The Correspondence of Isaac Newton*, 7 vols. (Cambridge, 1959-77), Vols. 1-3 ed. H. W. Turnbull, 3:336.

[12] Richard Bentley, "A Confutation of Atheism from the Origin and Frame of the World", in *A Defence of Natural and Revealed Religion: Being a Collection of the Sermons Preached at the Lecture founded by the Honourable Robert Boyle, Esq; (From the Year 1691 to 1732)*, ed. by Sampson Letsome and John Nicholl (London, 1739), 1:52-87. N. B. the title of this collection of lectures.

gument. These three sermons show the first adaptation of Newton's discoveries of the laws of the physical universe, especially the law of gravity, to the design argument.[13] But if Whiston's exposition of this Newtonian version of the design argument is not the first, it is one of the most important because of its thoroughness and especially because it clearly reveals the Newtonian concern with establishing the relationship between scientific reason and the historical world.

Whiston's first order of business in the *Astronomical Principles* is to set out as clearly and simply as possible the Newtonian account of the "System of

---

[13] See, for example, Robert H. Hurlbutt, *Hume, Newton, and the Design Argument* (Lincoln, Nebraska, 1965), p. 58; Louis Trenchard More, *Isaac Newton, A Biography* (New York, 1934), p. 375; Henry Guerlac and M. C. Jacob, "Bentley, Newton, and Providence," *Journal of the History of Ideas* 30 (July-Sept., 1969), p. 311.

the Universe". Whiston places great emphasis in his description of this system on the role of "universal attraction", or gravity:

> (1.) There is an universal Power of Gravity acting in the whole System; whereby every Body, and part of a Body, Attracts and is Attracted by every other Body and Part of a Body, through the whole System. (2.) This Power of Gravity is greater in greater Bodies, and lesser in lesser; and that in the proportion of such their magnitude. (3.) It is also greater when the Bodies are nearer, and lesser when they are farther off, and that in the exact proportion of such their nearness. (4.) This Power is the same in all Places, and at all times, and with regard to all bodies whatsoever: (5.) This Power is entirely immechanical, and beyond the Abilities of all material Agents whatsoever. [14]

---

[14] Whiston, *Astronomical Principles of Religion, Natural and Reveal'd* (London, 1717), p. 40.

Whiston "demonstrates" the truth of
this "Particular Account" of the system
of nature in ways familiar to readers
of Newton's *Principia*. However, Whiston
ultimately wishes to erect the design
argument, i.e., to infer the being and
nature of God on the basis of the order
and design exhibited in the whole frame
of physical nature. And to be able to
do this, Whiston must take great care
to show that the cause of the gravitational force whose operation he describes is not itself mechanical. Whiston
accordingly gives a quite detailed
"demonstration" of the fifth part of
the above quoted proposition (n. 14).
As an empiricist he points out that
this power does not exert its force
in the same way that a material agent
does and so is not itself material:

> This Power acts upon Bod-
> ies equally, when they
> are in the most violent
> Bodies with us; and the
> Celerity of the Comets
> in the Heavens, Geomet-

> rically computed, do particularly shew. Now this is absolutely impossible; that any Mechanical Pressure, or Impulse from a Body, let its Motion be never so Swift, or its Pressure never so strong, should equally accelerate another Body, when at Rest, and when in Motion; it being a known Law of Mechanism, that a Body in Motion impels another at Rest with its whole Force; but one in Motion, with only the Excess of its own Velocity above the others; as is the most obvious also on the least Reflexion. 15)

Another obvious disparity between the way a material agent exerts force on a body and the way gravity does is that material agents must be physically present whereas gravity acts on bodies at a distance. Hence, Whiston concludes that the nature of this power is completely non-mechanical:

---

15) *Ibid.*, p. 45.

> By this Power Bodies act
> upon other Bodies *at a
> Distance*, nay at all Dis-
> tances whatsoever; that is,
> they *act where they are
> not:* Which is not only
> impossible for Bodies Me-
> chanically to do, but
> indeed is impossible for
> all Beings whatsoever to
> do, either Mechanically
> or Immechanically ...   16)

After these descriptions and observations Whiston reaches finally his central point in Part VI concerning the "Important Principles of Natural Religion demonstrated from the foregoing *Observation*." Having described in detail the Newtonian system and the necessarily non-mechanical cause of the universal power of gravity which operates within that system, Whiston infers the being and the providential nature of God:

> We hence learn the *Being
> of God*, the first Intelli-
> gent Cause and Author, the

---

16) *Ibid.*, pp. 45-46.

just Owner and Possessor,
the Supreme Lord and
Governor, the constant
Preserver and Disposer
of all Things. This
Foundation of all Reli-
gion, the Belief of a Su-
preme Deity, is the
first, the most natural
and obvious Deduction
of Human Reason, even
from the Contemplation
of the most common and
ordinary Appearances of
Nature; from the Growth
of every Plant, and the
Succession of every Season,
and the general View of
Every Heavenly Body, and
every Creature about us.
And there have certainly
been no Nations or Peoples,
of the usual Capacities of
Mankind, but have ever
drawn this Consequence
in all Ages of the World.
So that if this instance
be not the *Voice of Nature*
itself, we shall be at a
loss to find other Truths,
requiring any Reasoning at
all, that can deserve to
be so stiled. And no wonder,
since the Argument is the
very same by which, from
the Contemplation of a
Building, we infer a Builder; and from the Elegancy

> and Usefulness of each
> Part, we gather he was a
> skilful Architect; or by
> which from the View of
> a Piece of Clockwork, we
> conclude the Being of the
> Clockmaker; and from the
> many regular Motions
> therein, we believe that
> he was a curious Artificer. 17)

Thus the whole Newtonian frame of nature and especially the immaterial power of gravity exhibit systematic order and design implying a designer identified by Whiston as God.[18] The Newtonian system is but the latest and best example of design and order which leads to this conclusion. And its very nature contains a feature which allows us to infer the providential nature of the Supreme Designer. In the 1706 *Opticks*, Newton formulates his views

---

17) *Ibid.*, p. 106.

18) Newton makes the same identification of the design argument's divine designer and the One God, the "Lord of Lords", in his own statement of the design inference in the "General Scholium".

that the world and its motion is in a
perpetual, although gradual, state of
decay due to the force of gravity causing all matter to collapse inwardly
to the "middlemost" of the universe.
The system of the universe therefore
requires, once in a great while, a small
adjustment.[19] Newton very much wants
a continuing place for God in this mechanical-geometrical universe and conceives of him not only as the generally
provident creator of the frame of nature but also as its specially provident preserver. In a rather startling
text, buried in his correspondence,
Newton indicates that it is precisely
a specially provident role which he be-

---

[19] Although the motion of the sun and the planets would remain relatively steady over "many Ages", irregularities in their orbits caused by gravitational attraction toward the center "will be apt to increase, till this System wants a Reformation ..." See Isaac Newton, *Opticks* (London, 1730), p. 402. This text is from Query 23 in the 1706 Latin *Optice* which becomes Query 31 in the 1718 translation and in successive editions.

lieves God fulfills by preventing the inward collapse of the whole system as the result of the power of gravity. God is both a generally provident creator of the system and a specially provident preserver of the system who protects it from itself. Newton observes that "a continual miracle is needed to prevent the sun and fixed stars from rushing together through gravity".[20]

In the *Astronomical Principles* Whiston expands on Newton's scattered hints. As a result of the power of gravity operating throughout the solar system, Whiston forthrightly states that

> it follows, that the
> several Systems, with
> their several Fixed
> Stars or Suns, do nat-
> urally and constantly...
> approach nearer and
> nearer to the common
> Center of their Gravity;
> and that in a sufficient
> Number of Years, they
> will actually meet
> in the same common

---

[20] Newton, *The Correspondence,* 3:336.

> Center, to the utter
> Destruction of the
> whole Universe. 21)

If the power of gravity were suspended for an instant, then the whole system would immediately dissolve. But the system remains in place despite the tendency of the law of gravity to collapse it inwardly. Whiston thus concludes that

> From the foregoing *System* we learn that God, the Creator of the World, does also exercise a continual *Providence* over it, and does interpose his general, immechanical immediate *Power*, which we call the *Power of Gravity* ... and without which all this beautiful System would fall to Pieces, and dissolve into Atoms. 22)

---

21) Whiston, *Astronomical Principles*, pp. 88-89.

22) *Ibid.*, p. 111. See also p. 82.

b. Synthesizing "Religion, Natural and Reveal'd" -- Forging the Initial Link Between Scientific Reason and the Historical World

The new Newtonian design argument was immediately accepted even among deists and freethinkers. Some time after Bentley's first presentation of the Newtonian version of the design argument in the 1692 Boyle Lectures, the noted deist Matthew Tindal states that mankind now has definite proof of the existence of a deity "from the marks we discern in the laws of the universe and its government".[23] But the very triumph of this first statement of the Newtonian design argument precipitates the attempt on the part of these deistic thinkers to separate natural religion, and its corollary in the design argument, from the revealed religion of historical scripture. Ironically, the very success of

---

[23] Matthew Tindal, *Christianity as Old as Creation* (London, 1735), p. 191.

Newton's "wonderful discoveries" in establishing natural religion causes the "Deist Sect" to turn their destructive scepticism to the task of subverting revelation. According to Whiston, his friend Bentley

> demonstrated the Being
> and Providence of God,
> from Sir *Isaac Newton's*
> wonderful discoveries,
> to such a degree of sat-
> isfaction, as to the
> scepticks or infidels
> themselves, that he
> informed me himself,
> of a club of such
> people, who had heard
> his sermons, and were
> asked by a friend of
> his, at his desire, what
> they had to say against
> them? They honestly
> owned that they did not
> know what to say. But
> added withal, what
> is this to the fable
> of *Jesus Christ*? Which
> made him say, that he
> doubted that he had done
> harm to Christianity by
> those sermons; as occa-
> sioning these scepticks
> or infidels to divert
> their denial of a God
> and a providence, from

> which they might be
> driven with great ease,
> to the picking up ob-
> jections against the
> Bible in general .... 24)

But, for the Newtonians, natural and revealed religion can never be split apart. The design argument is for them only the opening salvo in the defense of revealed religion. It is to demonstrate this point that Whiston writes his book and it is in his explanation of the connection between the two that its chief importance lies. Both natural religion and the historical religion of revelation are based on the "astronomical principles" of Newtonian science. Whiston says

> Since it has now pleased
> God, as we have seen,
> to discover many noble
> and important Truths to
> us, by the Light of Nature,
> and the System of the
> World; as also, he has
> long discovered many more
> noble and important Truths

---

[24] This anecdote is related in both Whiston's *Astronomical Principles,* p. 243, and his *Memoirs,* p. 93.

> by Revelation, in the
> Sacred Books; It cannot
> be now improper, to
> compare these two Divine
> Volumes, as I may well
> call them, together;
> in such Cases, I mean, of
> Revelation, as relate
> to the Natural World,
> and wherein we may be
> assisted the better to
> judge, by the Knowledge
> of the System of the
> Universe about us. For
> if those things con-
> tained in Scripture be
> true, and really deriv'd
> from the Author of Nature,
> we shall find them, in
> proper Cases, confirm'd
> by the System of the World;
> and the Frame of Nature
> will in some Degree, bear
> Witness to the Revelation. 25)

For Whiston the design argument becomes a part of a grand scheme of Biblical criticism and the mathematical astronomer becomes an exegete. Thus, as a Newtonian mathematician, Whiston shows how the two "Divine Volumes", scripture and

---

[25] Whiston, *Astronomical Principles*, p. 133.

science, confirm each other. The most prominent example is Whiston's interpretation of the Mosaic account of Creation, a theme first broached in his *A New Theory of the Earth*. Whiston interprets the story of the six days of creation in Genesis as the description of what an ordinary person would have reported seeing had one been present to witness the changes transpiring on the face of the earth. As such, the hexameron is a true "Historical Relation". But it is *not* a "Nice and Philosophick Account" of the sort which would satisfy a critical Newtonian.

For such a Newtonian description of, for example, the event described in Gen. 1:2 ("The earth was without form and void, and darkness was upon the face of the deep") one must look to the composition of comets. The nature of comets, or "worlds in confusion",[26] answers perfectly to both Moses' "Historical Relation" and to the "Nice and Philo-

---

[26] *Ibid.*, p. 23.

sophick Account" of a Newtonian scientist Whiston shows how the divine volume of Genesis is perfectly compatible with, and supported by, the divine volume of the best scientific data regarding comets. Genesis literally describes a comet, a "world in confusion" without form or void, a veritable chaos of particulate matter. Each facet of the Genesis creation story refers to a change in state in the comet's atmosphere, as its particles gradually settle out in accordance with their specific gravities, or as it shifts in its orbit around the sun. From the creation, through Noah's flood, to the inevitable final dissolution of this earth, Whiston relies on comets as a scientifically based explanation of what has happened or a scientific prediction of what will happen. But underlying this view is the crucial point that, for Newtonians such as Whiston, the revelation of historical scripture and the "*Voice of Nature* itself" are indeed different volumes, but ones which contain the same truth.

And what of cases where the two "Sacred Books", the "Divine Volumes" of nature and scripture, seem to disagree? We have only to compare them using the absolutely certain Newtonian system as the criterion of truth, i.e., the "Divine Volume" of nature is the standard by which the truth of the "Divine Volume" of historical revelation is to be judged. When divine revelation is found to be in conflict with the "Frame of Nature, which is now much better understood than in the Days of those Ancient Writers",[27] a forgery has been detected.

Whiston thus distinguishes between the level of certainty attainable in natural religion and that obtainable in revealed religion. The design argument, with its "noble Inferences", Whiston regards as certainly demonstrated because it is founded on the

---

[27] *Ibid.*, p. 134.

absolutely certain truth of the Newtonian system. In this, Whiston goes far beyond Newton who holds that the conclusions of both astronomy and natural religion are not on the same level of certainty as purely mathematical demonstrations.

But Whiston was the first to present Newtonian Science to non-mathematicians in the packed lectures given at Button's coffee house in Covent Garden. Accordingly, Whiston's view of the authority of the Newtonian system is far more widespread than Newton's own view. Pope's famous couplet accurately reflects the prevailing understanding of the certainty of Newtonian science and it is well to remember that Pope learns the Newtonian system by attending Whiston's lectures at Button's. Pope writes

> Nature and Nature's laws
> lay hid in night:
>
> God said, Let Newton be!
> and all was light.

Whiston does not hesitate to use this new criterion as a means for bringing historical revelation into line with scientific reason. Not surprisingly, Whiston ridicules "Popish astronomers" who reverse this order and make scripture the criterion of truth in judging scientific claims.[28] Whiston, the Newtonian scientist and Biblical critic, always is assured that the two "Divine Volumes" of natural religion and revealed religion are absolutely harmonious. If any minor synchronizations are required to combat the pernicious doctrines of Catholicism or the injudicious misinterpretations of the deists, the Newtonian scientist alone possesses the standard for doing so.

Whiston thus plainly acknowledges the theoretical superiority of the level of certainty of science against that of revelation. But this creates for Whiston and the majority of other early Newtonians no tension or "warfare" be-

---

[28] *Ibid.*, p. 39.

tween their science (and their natural religion) and their revealed religion. Such a warfare can arise only when the Newtonian system is really contradicted by revelation and that is impossible. Indeed, for Whiston the opposite side of the coin is of far greater importance. Where the "Divine Volumes" are found to agree, the revelation is substantially confirmed. As Whiston puts this point, the plain agreement of Nature, "or certain History", with Scripture

> will ... be a mighty Evidence for the Truth; and Uncorruptness of those Scriptures; and this even in general, as to such other Contents of the same, as can no way come under the like Methods of Examinations. If I am once fully satisfy'd, that a Witness is Upright and Honest, even in several Points where there was the greatest Suspicion as to his Sincerity, he will deserve the better Credit in other Cases, even where no corroborating Evidence

> can be alleg'd for his
> Justification. To this
> kind of Evidence then
> do I appeal on behalf
> of those Sacred Writings. 29)

In sum, Whiston takes the allegedly certain testimony of the results of Newton's scientific investigations to corroborate the probable testimony of historical revelation and then extends this argument to show that because the Bible has proven to be consistent with the true system of nature in every examined case, the general reliability of scripture in other cases is established. Far from utilizing the superior level of certainty of scientific results to prise science and revelation apart, Whiston sees science as a corroborating character witness to the overall reliability of historical revelation. Furthermore, he utilizes this method *against* precisely the group engaged in challenging the "fables" about Jesus Christ,

---

29) *Ibid.*, p. 134.

the deists. Thus, the transformation from Newtonian mathematician and natural scientist into Biblical critic and exegete, together with the linkage this supposes between scientific reason and history, is nearly complete.

c. The Final Link between Natural and Revealed Religion -- Newton, Whiston, and the testimony of the Ancients

Just before his famous methodological statement that "the main business of natural philosophy is to argue from phaenomena without feigning hypotheses, and to deduce causes from effects, till we come to the very first cause", Newton remarks that the "oldest and most celebrated philosophers of *Greece* and *Phoenicia*" share his own scientific theories.[30] They, too, reject a "dense fluid" medium (i.e., a plenum). They,

---

[30] Newton, *Opticks*, in *Opera Quae Exstant Omnia*, ed. S. Horsley, 5 vols. (London, 1782), 4:237.

too, affirm the existence of a vacuum
and atoms. And they, too, affirm the
operation of gravity between these
atomic bits of matter. Newton then
appends a very important but easily
missed phrase that these same ancient
philosophers "tacitly" attribute the
cause of this power of gravity to
something other than dense matter.[31]
The one crucial element in the Newtonian design argument, namely that the
cause of gravity is "entirely immechanical", is, according to Newton,
tacitly espoused by these ancient philosophers.

Thanks to the work of McGuire and
Rattansi, it is now evident that modern
scholars can no longer afford to consider such references to ancient philosophers as "merely literary embellishments to a scientific work".[32] As
they show in their analysis of the so-

---

[31] *Ibid.*

[32] McGuire and Rattansi, "Newton and the 'Pipes of Pan'," p. 108.

called "classical scholia" to an unpublished edition of the *Principia*, it is certain that Newton believes that such references to the "anticipations" by thinkers from antiquity form an important part of his overall philosophy.

This component of Newton's thinking and his own placement of himself in the "prisca sapientia" tradition is directly traceable only in the vast Newton manuscripts. His published works contain only the most truncated versions of these beliefs, such as the very abbreviated text in the *Optics*.

Once again, Whiston's *Astronomical Principles of Religion, Natural and Reveal'd* shed much light on Newton's views. For Whiston, the Newtonian design theology he has so laboriously detailed in the first seven parts, including the certain inference to the existence of God in natural religion and the consequently probable confirmation of historical revelation, is revealed in Pt. VIII to be nothing

less than "the common Voice of Nature
and Reason, from the Testimonies of the
most considerable Persons in all Ages."[33]
In this section of nearly a hundred pages,
the book's longest single chapter,
Whiston ransacks the writings of classical antiquity, both sacred and profane, for corroboration of the newly
established harmony between the certain
Newtonian design argument and the revelation of historical scripture. Whiston
does at length what Newton does in brief
in the unpublished "classical scholia"
and more briefly still in passages from
the *Optics* and General Scholium. Whiston
marshals his texts as a lawyer marshals
his witnesses to show that revelation
speaks with the "Common Voice of Nature
and Reason" and that the certain demonstrations of natural religion have
always been in perfect accord with the
truth of Biblical revelation as the
"*Testimonies* of the most considerable

---

[33] Whiston, *Astronomical Principles*, p. 157.

Persons in all Ages" show.

After several standard Biblical texts which declare the design argument,[34] Whiston quotes an interesting passage from Clement of Alexandria's *Recognitions*,[35] before launching into a long series of quotations from Ralph Cudworth's *True Intellectual System of the Universe* which emphasize the design

---

[34] For example, Psalm 19:1. "The Heavens declare the Glory of God: And the Firmament sheweth his Handy-work."

[35] The Alexandrian Christians, and especially Clement, utilize the *prisca* tradition to convert pagans. According to this tradition, a pristine knowledge of a single religious/scientific truth flowed into the Greek world from an Hebraic fountainhead; in the course of time this single truth fragmented and became corrupted. Early Christians such as Clement naturally appealed to this tradition in their missionary efforts. In Clement's *Stromata*, for example, he claims that Greek thinkers borrow their science and philosophy from the Hebrew prophets. See Eduard Zeller, *A History of Greek Philosophy*, trans. S. F. Alleyne (London, 1889), Vol. 1, pp.28. Cited in McGuire and Rattansi, "Newton and the 'Pipes of Pan'", pp.128-9.

argument in the writings of the "ancient Heathen writers." Whiston quotes Orpheus' song about how heaven, earth, and seas were framed out from the ancient "chaos".[36] For all Orpheus' polytheism, Whiston (and Cudworth) believe that Orpheus

> declared also in his Explication, that there was a certain incomprehensible Being, which was the Highest and Oldest of all Things, and the Maker of All, and who produceth all out of nothing into Being, whether visible or invisible. [37]

Passing from Orpheus to the Greek philosophers, Whiston cites passages

---

[36] Whiston, *Astronomical Principles,* p. 194. Whiston here states that these ancient instances of the design argument are "generally taken from the very Learned *Dr. Cudworth's Intellectual System of the Universe* and that nearly as he has translated them ..."

[37] *Ibid.,* pp. 194-95.

from Thales,[38] Pythagoras,[39] Heraclitus,[40] Zeno of Elea,[41] Socrates,[42] to the effect that the world of matter is created by an immechanical, incorporeal substance, i.e. God. Whiston believes that the secular nature of this testimony only makes it more believable. He states, "I have omitted most of the Christian Writers, as here of less Force, and without Number; excepting a very few of the most eminent of our modern Philosophers; who were of the Laity also; and so on all Accounts truly unexceptionable Witness in this Case."[44]

---

[38] *Ibid.*, p. 195.
[39] *Ibid.*, pp. 195-96.
[40] *Ibid.*, p. 198.
[41] *Ibid.*, p. 201.
[42] *Ibid.*, pp. 202-04.
[43] *Ibid.*, pp. 201-02.
[44] *Ibid.*, p. 194.

These testimonies are not to support further the Newtonian natural philosophy and its attendant design argument in natural religion. For Whiston and most people in this era, if not for Newton himself, these were established as certain. Relying heavily on Cudworth who also influenced Newton,[45] Whiston is seeking historical confirmation, rather than confirmation from the "Divine Volume" of nature, that the deity inferred in the design argument is the deity of Biblical revelation. In further cementing the connection between natural and revealed religion in this way, Whiston is carrying out the Newtonian program only hinted at by Newton.

---

[45] McGuire and Rattansi, "Newton and the 'Pipes of Pan'," pp. 132-33.

d. The Millennium and the Newtonian vision of Hell

In the same way that Whiston synthesizes the new scientific truth of Newton's system with the creation story in revelation so he labors to harmonize Newtonian science with the revealed knowledge of an imminent dissolution of this earth. Whiston clearly believes that he is extending the Newtonian empirical method into the realm of revealed religion in his many books which purport to show how Jesus fulfills Old Testament prophetic predictions.[46] Whiston

---

[46] See especially Whiston's own Boyle Lectures, *The Accomplishment of Scripture Prophecy* (London, 1708). Whiston claims that the topic he chose for his lectures was "Sir *Isaac Newton's* original suggestion." See Whiston, *Memoirs,* p.98. This ties in with Guerlac's and Jacob's contention that Newton was involved behind the scenes in selecting Bentley as the first

states

> Nor do I find that Mankind are usually influenc'd to change their Opinions by any Thing so much, as by Matters of Fact and Experiment; either appealing to their own Senses now; or by the faithful Histories of such Facts and Experiments that appealed to the Senses of former Ages. And if once the Learned come to be as wise in Religious Matters, as they are now generally become in those that are Philosophical and Medical, and Judicial; if they will imitate the Royal Society, the College of Physicians, or the Judges in Courts of Justice; (which last I take to be the most satisfactory Determiners of Right and Wrong, the most impartial

---

Boyle Lecturer and encouraged "if he did not suggest" Bentley's topic. See Guerlac and Jacob, "Bentley, Newton, and Providence," p. 318. The picture which thus emerges shows Newton functioning as an *eminence grise* who utilizes the public platform of the Boyle Lectures to promulgate by proxy his own views concerning both natural and revealed religion.

and successful *Judges of Controversy* now in the World:) If they will lay no other Preliminaries down but our natural Notions, or the concurrent Sentiments of sober Person in all Ages and Countries; which we justly call the Law or *Religion of Nature* ... And if they will then proceed in their Enquiries about Reveal'd Religion, by real Evidence and Ancient Records, I verily believe, and that upon much Examination and Experience of my own, that the Variety of Opinions about those Matters now in the World, will gradually diminish; the Objections against the Bible will greatly wear off; and genuine Christianity, without either *Priestcraft* or *Laycraft,* will more and more take Place among Mankind. 47)

In his Boyle Lectures of 1708, a topic originally suggested to him by Newton, Whiston argues that revealed

---

47) Whiston, *A Supplement to the Literal Accomplishment of Scripture Prophecies* (London, 1725), pp. 5-6.

scripture accurately describes the
end of the physical world: the
"Divine Volume" of scripture predicts
when and how the "Divine Volume" of
nature will be terminated. The world,
according to Whiston's reading of
revealed scripture, will be destroyed
by a universal conflagration after
which Christ will return.[48] Naturally, "the very latest and best of
all" scientific discoveries support
Whiston's interpretation of revelation.[49] Newton's discovery of the
operation of "immechanical", hence
providential, gravity and his own
vigorous efforts rightly to understand revelation Whiston sees as
part of the preparation for the
second coming.[50]

---

[48] Whiston, *The Accomplishment of Scripture Prophecy,* p. 10.

[49] *Ibid.,* p. 95.

[50] Whiston, *Historical Memoirs of the Life of Dr. Samuel Clarke,* 3rd ed., (London, 1748), pp. 74-75.

Following the second coming will be the last judgment. After judgment the resurrected souls of the just will inhabit the air of the earth's former orbit. The wicked, though, will be consigned to a punishment of alternate burning and freezing until they are annihilated. The mechanism for this punishment is a comet:

> Now this Description does in every Circumstance so exactly agree with the Nature of a Comet, ascending from the Hot Regions near the Sun, and going into the Cold Regions beyond *Saturn*, with its long smoaking Tail arising up from it, through its several Ages or Periods of revolving, and this in the Sight of all the Inhabitants of our Air, and of the rest of the System; that I cannot but think the Surface or Atmosphere of such a Comet to be that *Place of Torment* so terribly described in Scripture ...   51)

---

51) Whiston, *Astronomical Principles*,

No text more fully reveals the
extent to which Newtonian scientific principles complement revelation.
In Whiston, science and prophetic
history, past and future, march hand
in hand. Science is certain and
gives probable confirmation to histor-

---

pp. 155-6. After the last judgement
the earth itself would be shifted
out of its orbit into the radi-
cal ellipse of a comet; it
would become a comet. See Whiston,
A New Theory of the Earth, pp. 282
and 449. Whiston's striking vision of the fate of the damned
took root in the imagination of
Herman Melville who used it with
good effect in his novel *Mardi:*
"But whither now? To the broiling
coast of Papua? That region of
sunstrokes, typhoons, and bitter
pulls after whales unobtainable.
Far worse. We were going, it
seemed, to illustrate the Whistonian
theory concerning the damned and
the comets;- hurried from equinoc-
tial heats to arctic frosts. To
be short, with the true fickleness
of his tribe, our skipper had
abandoned all thought of the Cacha-
lot. In desperation, he was bent
upon bobbing for the Right whale
on Nor'-West Coast and in the Bay of
Kamchatka." See Melville, *Mardi and a
Voyage Thither,* 2 vols. (New York, 1963),1:4.

ical revelation rightly interpreted. Because this history is confirmed we have good reason to believe that revealed predictions about the future, if rightly understood on the basis of sound historical scholarship, will also come to pass.

III.

Conclusion: Destroying the Newtonian Synthesis of "Religion, Natural and Reveal'd"

Newton and Whiston accept the Bible as fundamentally accurate history which can be made clear and rational by examining it in the light of the "Divine Volume" of nature. For them the scientific theism based on the Newtonian system is inextricably linked with the historical revelation. Not only does the one "Divine Volume" of nature confirm the

"Divine Volume" of revelation, as in
the explication of the Mosaic history of creation, but the "*Testimonies
of the most considerable Persons in
all Ages*" also support the veracity
of Biblical history regarding the
being and nature of the one God.
The real importance of this view
is the highly probable guarantee
it provides for believing that the
future history of the world and of
mankind will conform to the as yet
unfulfilled Biblical prophetic history. As Whiston shows, this view
of the synthesis between Newtonian
science and the prophetic history
of the Bible is an extremely important part of the Newtonian philosophy as a whole. But until quite
recently even people who take Newton's version of the design argument
as an important element in his
thought ignore the Newtonian emphasis
on connecting divine history with
the new science. Thus Hurlbutt makes

the following oversimplification:

> Where the Deistical point
> of departure was primarily
> philosophical and historical
> on both the positive and
> negative sides, the Newtoni-
> ans based their thinking
> in scientific ideas and
> discoveries, and tried
> insofar as it is possible
> to have it accord with the
> observational methods and
> techniques so closely
> related to those
> discoveries. 52)

If it seems odd to include Newton and followers such as Whiston with modern day fundamentalists who preach the imminence of the "world to come," it is because we have a completely different outlook on Biblical history and on science. Today, we see these as two completely separate endeavors of which science is by far the more respectable. While there are historians of the Bible today in modern secular universities, there are no

---

52) Hurlbutt, *Hume, Newton, and the Design Argument*, p. 77.

historians who take the Bible as past and future historical truth confirmed by the Department of Astrophysics. Through the eighteenth century, however, the Newtonian synthesis was maintained by such scientists as David Hartley and Joseph Priestley and by such historians as William Robertson.[53]

But the predominant trend of historical thinking in the eighteenth century is secular and it is by the emerging secular historians of the Enlightenment that we in the modern era have been chiefly influenced.

By far the most subtle of these new, secular-minded social scientists is David Hume. It is one more irony in a career filled with ironies to

---

[53] Hartley, *Observations on Man* (London, 1752); Priestley, *Letters to a Philosophical Unbeliever* (London, 1780); Robertson, *History of America* in *Works of William Robertson* (London, 1827), Vol. 6.

note how many elements of the Newtonian synthesis of Biblical history and natural science that Hume, who professes the desire to explain the hidden springs and principles of human nature just as Newton explains physical laws, attempts to annihilate.[54] The "great infidel" is, of course, most famous for his direct assault on the logic of the design argument in his *Dialogues Concerning Natural Religion*. But Hume takes as much care to sever natural from revealed historical religion as he does to undermine the mode of analogical reasoning employed in the design argument.

Hume is himself an historian of

---

[54] David Hume, *Enquiry Concerning Human Understanding and Concerning the Principals of Morals,* ed. L. A. Selby-Bigge (Oxford, 1975), p. 14. The actual reference to Newton's influence on Hume is in Section I of *An Enquiry Concerning Human Understanding.*

consequence.[55] But Hume also restricts human knowledge to human experience. Even historical knowledge must be measured by the yardstick of our own experience because history is but an extension of our own experience:

> And indeed, if we consider
> the shortness of human
> life, and our limited knowledge,
> even of what passes in our
> own time, we must be sen-
> sible that we should be for
> ever children in understand-
> ing, were it not for this
> invention which extends
> our experience to all
> past ages and to the most
> distant nations. [56]

---

[55] Hume's *History of England* was the most important history of England produced in the eighteenth century. See the introductory essays by Richard H. Popkin and David Fate Norton in *David Hume: Philosophical Historian*, ed. with Introductory Essays, by David Fate Norton and Richard H. Popkin (Indianapolis, New York, 1965.)

[56] David Hume, "Of the Study of History," in *David Hume: Philosophical Historian*, p. 38.

So long as we confine our reasoning to that with which we have experience, whether in speculations regarding history, trade, morals, politics, or criticism, we operate within the just limits of our powers of understanding.

> But when we look beyond human affairs and the properties of the surrounding bodies; *When we carry our speculations into the two eternities, before and after the present state of things;* into the creation and formation of the universe; the existence and properties of one universal spirit, existing without beginning and without end, omnipotent, omniscient, immutable, infinite, and incomprehensible: We must be far removed from the smallest tendency to scepticism not to be apprehensive that we have here got quite beyond the reach of our faculties. 57)

---

57) David Hume, *Dialogues Concerning Natural Religion*, ed. Norman Kemp Smith (Indianapolis, New York, 1947), pp. 134-35. Emphasis added.

In theological-scientific inquiries regarding the Newtonian mechanism for the creation of this world as recounted in Moses' prophetic history and the ultimate dissolution of this earth as prophesied in unfulfilled prophetic history, we have transcended the limits of what we can justifiably claim to know. Hume states (speaking through Philo in the *Dialogues*):

> The first step, which we make, leads us on for ever. It were, therefore, wise in us, to limit all our enquiries to the present world, without looking farther. No satisfaction can ever be attained by these speculations, which so far exceed the narrow bounds of human understanding. 58)

Concerning Newton's own attempt to explicate prophetic history in his

---

58) *Ibid.*, p. 162.

Observations upon the Prophecies of Holy Writ, particularly the Prophecies of Daniel and the Apocalypse of St. John (1733), Hume magnanimously remarks that such work only reveals the ignorance of the preceding generation and that we "never should pronounce the folly of an individual, from his admitting popular errors, consecrated by the appearance of religion."[59]

The main theme of Hume's essay "Of Miracles" is to show the plausibility of some kinds of historical accounts and the total implausibility of *any* historical account which does not square with our own experience and common sense. To believe in the history recorded in scripture of miraculous events, i.e., events which contravene the laws of nature, requires an inward, non-rational personal miracle, an event which Hume

---

[59] Hume, *The History of England*, in *David Hume: Philosophical Historian*, p. 266.

clearly feels is impossible for a rational man in his own enlightened age.[60] But, as we have seen, the chief reason the Newtonians are concerned with correctly interpreted prophetic history is for what it shows about the course of future history in unfulfilled prophetic predictions. Significantly, Hume's essay "Of Miracles" applies the canons of common sense and ordinary experience to prophecies as well as to miracles. Hume makes it explicit that prophecies are but a sub-class of miracles and, hence, that the scripture histories regarding them are simply too bizarre, too extraordinary, too unusual to be believed.[61] So much for the Newtonian view that science confirms such historical accounts and thus helps to guarantee

---

[60] Hume, *Enquiry Concerning Human Understanding*, Section X, p. 130.

[61] *Ibid.*, pp. 130-31.

our belief in the fulfillment of future prophetic history. Hume simply sets aside this aspect of Newtonianism.

Having severed natural from revealed religion, Hume proceeds to examine the design argument wholly on its logical merits. Hume concentrates his attack on the method of analogy proponets of the design argument employ to infer the nature of God. Whiston, to take one clear example, believes that this method of reasoning is sound:

> And if there be any Deductions of Human Reason which are easier and more obvious than the rest, this Way of Arguing, which we have already used, from the *House* to the *Architect*; from the *Clock* to the *Clockmaker*; from the *Ship* to the *Shipbuilder*; and from a *noble, large,* well-contriv'd and well-proportion'd, and most beautiful House, or Clock, or Ship, to the *excellent* Architect, the *skilful* Clockmaker, the *sagacious* Shipbuilder; this is such clear natural

> obvious, sure Reasoning,
> that we even at first make
> use of it in Childhood,
> and find it as clear,
> natural, obvious, and
> sure in our elder Age;
> without occasion for a
> tutor to instruct us
> in it at first, or for
> a Logician to improve
> us in it afterwards.   62)

Hume clearly agrees with Whiston about the natural psychological inevitability of this kind of argument; in fact, this is its sole strength as an argument for Hume. However, we cannot, according to Hume, deduce from what we know from experience anything positive about the nature of the creator. Indulging in such speculation carries us quite beyond the natural limits of our experience and, hence, of our knowledge. We simply haven't experience of world-making or universe-building by God and so cannot use this small earth, which is but a small part of a very

---

62) Whiston, *Astronomical Principles*, p. 255.

large whole, as the basis for an
inference regarding the nature of
its maker (or makers).[63] We can
certainly infer the being and quality of an architect from a particular
example of his work, but this case
is simply not analogous to inferring
an omniscient, omnipotent deity
from our world system. Hume explicitly mentions Whiston's design argument trademark, the theory of the
cometary origin of the earth and
rejects it as "beyond our faculties."[64]
Hume concludes his book with the famous claim that from the nature of

---

[63] Hume, *Dialogues*, pp. 148-49.

[64] *Ibid.*, p. 177. Hume says, "A Comet", for all we may know, could be "the seed of a world; and after it has been fully ripened, by passing from sun to sun, and star to star, it is at last tossed into the unformed elements, which everywhere surround this universe, and immediately sprouts up a new system." Whiston, as we have seen, theorizes that comets play just usch a role in God's providential system.

the world around us, we are justified
only in the admittedly ambiguous inference that "*the cause or causes of
the order in the universe probably
bear some remote analogy to human
intelligence.*"[65]

In sum, Hume severs the design
argument from historical revelation;
it is begging the question to assume
that both or either nature or revelation are "Divine Volumes" on the basis of what we know from experience.
Next, Hume shows how empty and remote is the claim of Newtonian design
theology once divorced from prophetic
history. The Design Argument reveals
not the God of scriptural revelation;
it is able to infer only the vaguest
abstraction.

How different this modern paganism
is from the views of those first modern scientist-theologians who immediately adapt the Newtonian system
to the task of confirming Christian

---

[65] *Ibid.*, p. 227.

revelation and thereby fleshing out
the nature of their God. As Whiston
puts it:

> The Reveal'd Religion
> of the Jews and Chris-
> tians lays the Law of Na-
> ture for its Foundation,
> and all along supports
> and assists Natural
> Religion; as every
> True Revelation
> ought to do.   66)

It is Hume who first perceives
that to proceed on this assumption
is truly to follow the primrose
path. But that fact should not
obscure for us that this naive
symbiosis between scientific reason and historical revelation is
in fact as much a part of the Newtonian program as the law of
gravity. Whiston reveals this
simple truth better than anyone

---

66) Whiston, *Astronomical Principles*, p. 259.

of his like-minded contemporaries including Newton.[67]

James E. Force

The University of Kentucky

---

[67] I should like to express my appreciation to the NEH, without whose assistance this paper could not have been written.

ASTRONOMICAL
Principles of Religion,
NATURAL and REVEAL'D.

In NINE Parts:

I. *Lemmata*; or the known Laws of Matter and Motion.
II. A particular *Account* of the System of the Universe.
III. The *Truth* of that System briefly Demonstrated.
IV. Certain *Observations* drawn from that System.
V. Probable *Conjectures* of the Nature and Uses of the several Celestial Bodies contained in the same System.
VI. Important Principles of NATURAL RELIGION Demonstrated from the foregoing *Observations*.
VII. Important Principles of DIVINE REVELATION Confirm'd from the foregoing *Conjectures*.
VIII. Such Inferences shewn to be the common Voice of Nature and Reason, from the *Testimonies* of the most considerable Persons in all Ages.
IX. A *Recapitulation* of the Whole: With a Large and Serious *Address* to all, especially to the *Scepticks* and *Unbelievers* of our Age.

Together with
A PREFACE,
Of the *Temper of Mind* necessary for the Discovery of *Divine Truth*; and of the *Degree of Evidence* that ought to be expected in *Divine Matters*.

By *WILLIAM WHISTON*, M. A.
Sometime Professor of the Mathematicks in the University of CAMBRIDGE.

LONDON: Printed for J. SENEX at the *Globe* in *Salisbury-Court*, and W. TAYLOR at the *Ship* in *Pater-noster-Row*, 1717.

To the Illustrious

Sir *Isaac Newton*,

PRESIDENT,

And to the rest of the

COUNCIL and MEMBERS

OF THE

ROYAL SOCIETY;

THESE

ASTRONOMICAL

𝕻rinciples of 𝕽eligion

NATURAL and REVEAL'D,

ARE

Most Humbly DEDICATED

BY

March 25. 1717.

The AUTHOR.

ERRATA.

PAGE 15. Line 2. read *about* 45000. p. 16. l. 8. r. 200,000.000. p. 25. l. 13. r. 47″. p. 50. l. 20. *Marg.* r. *Fig.* V. p. 135. l. 30, 31. r. *Mankind: For all this*, &c. [Vid. Errat. reliq. in calce, p 304.]

PREFACE.

BEFORE *I come to treat of this Noble Subject,* The Astronomical System of the Universe, *with its wonderful Consequences, as it is now discovered to us by the good Providence of God, and the laborious Searches of this and the last Age; and chiefly, by the surprizing Sagacity and Penetration of the Illustrious Sir* Isaac Newton; *I think it proper to premise two Enquiries, as of great Consequence in way of Preparation to the receiving real Advantage by this Treatise; and without Satisfaction wherein, all such Discourses will be of little Effect with many of its Perusers: I mean the Enquiry about that*

Temper

PREFACE.

Temper of Mind, *which is necessary for the Discovery of Divine Truth; and the Enquiry into that Degree of Evidence, that ought to be expected in Divine Matters. For, if all our Knowledge be derived from God, and if it has pleased God to require a certain Degree of Probity, Seriousness, Impartiality, and Humility of Mind; together with hearty Prayers to him for his Direction, Blessing, and Assistance; and a proper Submission to him, before he will communicate his Truths to Men; I mean, at least, communicate the same so as shall make a due Impression upon their Minds, and turn to their real Profit and Edification, to their true Improvement in Virtue and Happiness: And if Men at any time come to the Examination either of the Works or Word of God, without that Temper of Mind, and without those Addresses for his Aid, and Submission to his Will, which He has determined shall be the Conditions of his Communications to them; especially if they come with the contrary Dispositions, with a Wicked, Partial, Proud, and Ludicrous Temper, and with an utter Disregard to God, his Providence, Worship, and Revelation; all their Researches will come to nothing. If, I say, this be the Case, as to Divine Knowledge, as I believe it is, it cannot but be highly necessary for us all to consider of this Matter beforehand, and to endeavour after the proper*

Qualifi-

Qualifications, before we set our selves about the main Enquiries themselves. If it has also pleased God to expect from us some more Deference and Regard for him, than for our poor fallible Fellow-Creatures here below; and to claim our Belief and Obedience, upon plain external Evidence, That certain Doctrines or Duties are derived from him, without our being always let into the Secrets of his Government, or acquainted with the Reasons of his Conduct; and also to expect that this plain external Evidence be treated, as it is in all the other Cases of Human Determinations and Judgments; I mean, that it be submitted to, and acquiesced in, when it appears to be such, as in all other Cases would be allowed to be satisfactory, and plainly superior to what is alledged to the contrary: If, I say, this also be the Case as to Divine Knowledge, as I believe it is; It will be very proper for us all to consider of this Matter before-hand also; that so we may not be afterward disappointed, when in our future Progress we do not always find that irresistible and over-bearing Degree of Evidence for certain Divine Truths, which in such Cases is not to be had; which in truth is almost peculiar to the Mathematicks; and the Expectation of which is so common, tho' unjust, a Pretence for Infidelity among us.

As to the former of these Enquiries, or that Temper of Mind *which is necessary for the Discovery of Divine Truth; it can certainly be no other than what the Light of Nature, and the Consciences of Men influenc'd thereby, dictate to us; those, I mean, already intimated; such as Seriousness, Integrity, Impartiality, and Prayer to God; with the faithful Belief, and ready Practice of such Truths and Duties, as we do all along discover to be the Word and Will of God; together with such a Modesty, or Resignation of Mind, as will rest satisfy'd in certain sublime Points, clearly above our Determination, with full Evidence that they are revealed by God, without always insisting, upon knowing the Reasons of the Divine Conduct therein immediately, before we will believe that Evidence. These are such Things as all honest and sober Men, who have naturally a Sense of Virtue and of God, in their Minds, must own their Obligation to. We all know, by the common Light of Nature, till we eclipse or corrupt it by our own Wickedness, That we are to deal with the utmost Fairness, Honesty, and Integrity in all, especially in Religious Matters; that we are to hearken to every Argument, and to consider every Testimony without Prejudice, or Byas, and ever to pronounce agreeably to our Convictions; that we are but Weak, Frail, Dependent*

PREFACE.

dent Creatures, all whose Faculties, and the Exercise of them, are deriv'd from God; that we ought therefore to exercise a due Modesty, and practise a due Submission of Mind in Divine Matters, particularly in the Search after the Nature, and Laws, and Providence of our great Creator: A Submission, I mean, not to Human, but to Divine Authority, when once it shall be authentickly made known to us: That the humble Addressing of our selves to God for his Aid, Direction, and Blessing on our Studies and Enquiries, is one plain Instance of such our Submission to Him; and that a ready Compliance with Divine Revelation, and a ready Obedience to the Divine Will, so far as we have clearly discover'd it, is another necessary Instance of the same humble Regard to the Divine Majesty. Nor indeed, can any one who comes to these Sacred Enquiries with the opposite Dispositions, of Dishonesty, Partiality, Pride, Buffoonry, Neglect of all Divine Worship, and Contempt of all Divine Revelation, and of all Divine Laws, expect, even by the Light of Nature, that God should be oblig'd to discover farther Divine Truths to him. Nor will a sober Person, duly sensible of the different States of Creator and Creature, imitate Simon Magus, *and his Followers, in the first Ages of the Gospel; and set up some Metaphysical Subtilties, or Captious Questions, about the Conduct of Providence, as sufficient to set aside the Evidence of confessed Miracles themselves; but will rather agree to that wise Aphorism laid down in the Law of* Moses, *and suppos'd all over the Bible;* That Secret Things belong unto the Lord; but Things that are revealed, to Us and to our Children, that we may do them. *Now in order to the making some Impressions upon Men in this Matter, and the convincing them, that All our Discoveries are to be derived from God; and that we are not to expect his Blessing upon our Enquiries, without the foregoing Qualifications, Devotions, and Obedience: Give me leave here, instead of my own farther Reasoning, to set down from the Ancient* Jewish *and* Christian *Writers, several Passages which seem to me very remarkable, and very pertinent to our present Purpose: Not now indeed, as supposing any of those Observations of Sacred Authority, but as very right in themselves; very agreeable to the Light of Nature; and very good Testimonies of the Sense of wise Men in the several ancient Ages of the World to this Purpose. And I chuse to do this the more largely here, because I think this Matter to be of very great Importance; because it seems to be now very little known or consider'd, at least very little practis'd, by several pretended Enquirers into Reveal'd Religion; and be-*

Deu. xxix. 29.

cause

PREFACE.

cause the Neglect hereof seems to me a main Occasion of the Scepticism and Infidelity of this Age.

The Lord spake unto *Moses*, saying; See, I have called by Name *Bezaleel*, the Son of *Uri*, the Son of *Hur*, of the Tribe of *Judah*: And I have filled him with the Spirit of God, in Wisdom, and in Understanding, and in Knowledge, &c. And in the Hearts of all that are wise-hearted, I have put Wisdom, &c. *Ex. xxxi. 1, 2, 3, 6.*

It shall come to pass, if thou wilt not hearken unto the Voice of the Lord thy God, to observe to do all his Commandments, and his Statutes, which I command thee this Day, that all these Curses shall come upon thee, and overtake thee: —— The Lord shall smite thee with Madness, and Blindness, and Astonishment of Heart; and thou shalt grope at Noon-day, as the Blind gropeth in Darkness. *Deut. xxviii. 15. v. 28, 29.*

The Lord hath not given you an Heart to perceive, and Eyes to see, and Ears to hear, unto this Day. *xxix. 4.*

Give thy Servant an Understanding Heart, to judge thy People; that I may discern between Good and Bad: For who is able to judge this thy so great a People? And the Speech pleased the Lord, *1 King. iii. 9, 10, 11, 12.*

A 4 that

that *Solomon* had asked this Thing. And God said unto him, becaufe thou haft asked this Thing; ———Haft asked for thy felf Underftanding to difcern Judgment; behold I have done according to thy Words: Lo, I have given thee a Wife and an Underftanding Heart; fo that there was none like thee before thee; neither after thee fhall any arife like unto thee. ———And all *Ifrael* heard of the Judgment which the King had judged; and they feared the King; for they faw that the Wifdom of God was in him, to do Judgment.

<small>v. 28.</small>

<small>Job xxxii. 7, 8.</small>
I faid, Days fhould fpeak; and Multitude of Years fhould teach Wifdom: But there is a Spirit in Man; and the Infpiration of the Almighty giveth them Underftanding.

<small>xxxiii. 12, 13.</small>
Behold in this thou art not juft; I will anfwer thee, that God is greater than Man. Why doft thou ftrive againft him? For he giveth not Account of any of his Matters.

<small>xxxiv. 31, 32.</small>
Surely it is meet to be faid unto God, I have born Chaftifement; I will not offend any more: That which I fee not, teach thou me; if I have done Iniquity, I will do no more.

<small>xxxvii. 5.</small>
God thundreth marvelloufly with his Voice: Great Things doth he which we canno comprehend. With

PREFACE.

With God is terrible Majesty: Touching the Almighty we cannot find him out: He is excellent in Power, and in Judgment, and in Plenty of Justice: He will not afflict. Men do therefore fear him: He respecteth not any that are wise of Heart. v. 22, 23, 24.

Who hath put Wisdom in the inward Parts? Or who hath given Understanding unto the Heart? xxxviii. 36.

Then *Job* answered the Lord and said; I know that thou canst do every Thing, and that no Thought can be with-holden from thee. Who is he that hideth Counsel without Knowledge? Therefore have I uttered that I understood not; Things too wonderful for me, which I knew nor. xlii. 1, 2, 3.

—I have heard of thee by the Hearing of the Ear, but now mine Eye seeth thee: Wherefore I abhor my self, and repent in Dust and Ashes. v. 5, 6.

The Meek will he guide in Judgment: The Meek will he teach his Way. Psal. xxv. 9.

The Secret of the Lord is with them that fear him; and he will shew them his Covenant. v. 14.

Thou through thy Commandments hast made me wiser than mine Enemies; for they are ever with me. I have more Understanding than all my Teachers, for thy Testimonies are my Meditation. I under- cxix. 98, 99, 100.

v. 104. understand more than the Ancients, because I keep thy Precepts. ——Through thy Precepts I get Understanding; therefore I hate every false way.

cxxxi. 1. Lord, my Heart is not haughty, nor mine Eyes lofty; neither do I exercise my self in great Matters, or in Things too high for me.

Prov. ii. 6. The Lord giveth Wisdom: Out of his Mouth cometh Knowledge and Understanding.

iii. 5, 6. Trust in the Lord with all thine Heart, and lean not to thine own Understanding. In all thy ways acknowledge him, and he shall direct thy Paths.

v. 32. The Froward is Abomination to the Lord: But his Secret is with the Righteous.

Eccles. ii. 26. God giveth to a Man that is good in his Sight, Wisdom, and Knowledge, and Joy.

iii. 11. God hath made every Thing Beautiful in his Time: Also he hath set the World in their Heart; so that no Man can find out the Work that God maketh, from the Beginning to the End.

viii. 17. Then I beheld all the Work of God that a Man cannot find out the Work that is done under the Sun; because though a Man labour to seek it out, yet he shall not find it: Yea further, though a wise Man

PREFACE.

Man think to know it, yet shall he not be able to find it.

As for these Four Children, God gave them Knowledge and Skill in all Learning and Wisdom. Dan. i. 17.

None of the Wicked shall understand, but the Wise shall understand. xii. 10.

Who is wise, and he shall understand these Things; prudent, and he shall know them: For the ways of the Lord are right, and the Just shall walk in them; but the Transgressors shall fall therein. Hof. xiv. 9.

And the Angel that was sent unto me, whose Name was *Uriel*, gave me an Answer, and said, Thy Heart hath gone too far in this World: And thinkest thou to comprehend the Way of the most High? 2 Esd. iv. 1, 2.

He said moreover unto me, Thine own Things, and such as are grown up with thee, canst thou not know; how should thy Vessel then be able to comprehend the way of the Highest? v. 10, 11.

They that dwell upon the Earth may understand nothing; but that which is upon the Earth: And he that dwelleth above the Heavens, may only understand the Things that are above the heighth of the Heavens, &c. v. 21.

Into

PREFACE.

Wifd. i. 41, &c. Into a malicious Soul Wifdom fhall not enter, nor dwell in the Body that is fubject unto Sin, &c.

ii. 21. Their own Wickednefs hath blinded them.

v. 22. As for the Myfteries of God, they know them not.

vii. 7. Wherefore I prayed, and Underftanding was given me: I called upon God, and the Spirit of Wifdom came to me:

v. 15, 16. It is God that leadeth unto Wifdom, and directeth the Wife. For in his Hand are both we and our Words; all Wifdom alfo, and Knowledge of Workmanfhip.

viii. 21. When I perceived that I could not otherwife obtain Wifdom, except God gave her me; (and that was a point of Wifdom alfo, to know whofe Gift fhe was,) I prayed unto the Lord, and befought him, and with my whole Heart I said :

ix. 4, 5, 6. Give mè Wifdom that fitteth by thy Throne, and reject me not from among thy Children. For I thy Servant, and Son of thine Handmaid, am a feeble Perfon, and of a fhort Time, and too young for the Underftanding of Judgment and Laws. For though a Man be never fo perfect among the Children of Men, yet

PREFACE.

yet if thy Wisdom be not with him, he shall be nothing regarded.

Hardly do we guess aright at Things that are upon Earth; and with Labour do we find the Things that are before us: But the Things that are in Heaven who hath searched out? *v. 16.*

All Wisdom cometh from the Lord, and is with him for ever. ——— She is with all Flesh according to his Gift; and he hath given her to them that love him. *Eccluf. i. 1. v. 10.*

If thou desire Wisdom, keep the Commandments, and the Lord shall give her unto thee. For the Fear of the Lord is Wisdom, and Instruction; and Faith and Meekness are his Delight. *v. 26, 27.*

Mysteries are revealed unto the Meek ——— Seek not out the Things that are too hard for thee; neither search the Things that are above thy Strength. But what is commanded thee think thereupon with Reverence: For it is not needful for thee to see the Things that are in Secret. *iii. 19, 21, 22.*

Let thy Mind be upon the Ordinances of the Lord, and meditate continually in his Commandments. He shall establish thine Heart, and give thee Wisdom at thine own Desire. *vi. 37.*

Wisdom, Knowledge, and Understanding of the Law, are of the Lord. Love, *xi. 15, 15.*

and

and the way of good Works, are from him. Error and Darkness had their Beginning together with Sinners.

xv. 7, 8. Foolish Men shall not attain unto Wisdom; and Sinners shall not see her. For she is far from Pride; and Men that are Lyars cannot remember her.

xxi. 11. He that keepeth the Law of the Lord getteth the Understanding thereof; and the Perfection of the Fear of the Lord is Wisdom.

xxxix. 24. As his Ways are plain unto the Holy, so are they Stumbling blocks unto the Wicked.

xliii. 33. The Lord hath made all Things, and to the Godly hath he given Wisdom.

John vii. 17. If any Man will do his Will, he shall know of the Doctrine whether it be of God, or whether I speak of my self.

Rom. xi. 33. O the Depth of the Riches both of the Wisdom and Knowledge of God! How unsearchable are his Judgments, and his Ways past finding out.

Jam. i. 5. If any of you lack Wisdom, let him ask of God, that giveth to all Men liberally, and upbraideth not, and it shall
v. 17. be given him. Every good Gift, and every perfect Gift, is from Above, and cometh down from the Father of Lights.

If

PREFACE.

If I once know that he is a Man of *Recog. ii. 4.* Probity, and unblamable in those Points of Duty wherein there can be no doubt but they are good; that is, if he be sober, if he be merciful, if he be just, if he be meek and humane; which no one can doubt to be virtuous and good Things; then it will, in all Probability, be reasonable to expect that to him that hath these excellent Virtues, that which is wanting to his Faith and Knowledge will be conferr'd; and that wherein his Life, which is so commendable in the rest, shall still appear blamable, it may be amended. But if he be involved and polluted in Sins, those I mean which are notoriously such; I must not then declare plainly to him the least Part of the conceal'd recondite Branches of Divine Knowledge: But rather, with great Boldness, deal with him that he must leave off his Sins, and amend his vicious Actions.

Whence 'tis very plain, that some do §. 16. oppose the Truth of the Religion of God, not because the Foundation of Faith does not seem to them certain; but because they are either involv'd in a Superabundance of Sins, or prepossess'd by their wicked Habits, or puff'd up by the Pride of their Heart; so that they do not believe

lieve even those Things which they think they see with their own Eyes.

§. 17. But now, because an innate Affection towards God our Creator, might seem sufficient for the Salvation of those that lov'd him, the Enemy studies to pervert the Affection of Men, and to render them Enemies and Ungrateful to their Creator, &c.

§ 18. We not only do enjoy God's Benefits, but by his Aid and Power it was that we came into Being, when we were not: whom also, if we please, we shall obtain from him, as our Reward, to be for ever in Happiness. To the End therefore, that Unbelievers may be distinguish'd from Believers, and the Pious from the Impious; the Evil one has Permission to make Use of these Arts, whereby every ones Affections towards their proper Parent may be tried, &c.

§. 19. Here therefore, that is while we continue in this present Life, where is the Place of Action, you ought to acknowledge the Will of God. For if any one has a mind to enquire after Things that cannot be found out, before he amends his Life, such an Enquiry is foolish, and will be to no Purpose. For Time is short; and the Judgment of God will be appointed on Account of Mens Actions, and not their Questions.

Queſtions. And therefore let us firſt of all make Enquiry what we are to Do, and after what manner it is to be done; that we may be thought worthy of Eternal Life. For if we ſpend this ſhort Time of Life in idle and unprofitable Queſtions, we ſhall certainly go to God empty, and deſtitute of good Works; at that Time, I mean, when the Judgment ſhall be appointed for our Works; for every Thing has its proper Time and Place. This is the Place, this the Time for Works: The World to come for Retribution. Left therefore we ſhould change the Order of Time and Place, and thereby be our own Hindrance; let our firſt Enquiry be what is *God's Righteouſneſs*; that like thoſe that are going a Journey, we may have a plentiful Proviſion for our Journey, that is, good Works; that ſo we may be able to arrive at the Kingdom of God, as at a very great City. For to thoſe who are well diſpos'd, God is manifeſted by thoſe Works of Nature which he has made, and is atteſted to by his own Creatures. Since therefore there ought to be no doubt concerning the Exiſtence of God, we are only to enquire about his Righteouſneſs, and his Kingdom. But if our Minds have an Inclination to put us on the Enquiry about ſecret and hidden Things, before we

§. 20.

a enquire

enquire after the Works of Righteousness, we ought to give an Account to our selves of this Procedure: For if we live well, and are thought worthy to obtain Salvation, we shall go to God Chaste and Pure, and be fill'd with the Holy Ghost; and shall know all such secret and hidden Things, without any Cavilling or Question; which at present, though any one should spend the entire Time of his Life in the Enquiry, he will be so far from finding them out, that he will bring himself into greater Errors; because he aims to arrive at the Haven of Life, without walking the way of Life.

iii. 19. Do but consider that Silence and Quietness with which all the People stand; and how, as you see, they are very Patient, and pay a great Honour to the Truths of God, even before they are instructed in them. For as to any greater Honour they have not yet learned that 'tis their Duty to pay it. For which Reason I have Hope, in the Mercy of God, that he will receive kindly this Religious Disposition of their Minds towards him; will afford the Reward of Victory to him that preaches the Truth; and will make manifest to them which of us is the Preacher of the Truth.

There-

PREFACE.

§. 37. Therefore is it necessary for Men to enquire whether they have it in their Power by seeking to find what is good; and when they have found it, to do it. For this is that for which they must be judg'd. As for what is beyond this, no one but a Prophet needs to know it: And with good Reason. For to what Purpose is it for Men to know after what Manner the World was made? Which yet would be necessary for us to know, if we were to undertake so artful a Piece of Work our selves. But now it is sufficient for us, and for our Worshipping God, to know that he made the World: But after what Manner he made it, we are not to enquire; because, as I said, 'tis not our Business to learn the Knowledge of that Art, as if we were to make somewhat like it. Nor are we to be called to Account for this, Why we did not learn after what Manner the World was made? But only for this, That we are ignorant of its Creator. But we shall know that God, the Creator of the World, is both Just and Good, if we seek after him in the Paths of Righteousness.

§. 52. God, who is the One and True God, resolv'd to prepare good and faithful Friends for his first-begotten Off-spring: But knowing they could not be Good, unless they had in their own Power that

Sense of the Things whereby they might become good; that they might be what they desired to be by their own Choice; and that otherwise they could not be really Good, if they were not such by Choice, but were forced to be such by a Necessity of Nature, he gave every one the free Power of his own Will; that he might be able to be such an one as he desired to be, &c.

See also ix. §. 4, 5, 6, 8.

§. 58.

As God has plac'd the Compass of Heaven above the Mountains, and the Earth, so has he covered the Truth with the Vail of his Charity; that he only might attain to it, who would first knock at the Gate of the Divine Love.

See iv. §. 4, 5. viii. §. 52, 53, 56, 58, 59.

viii. 61, 62.

Men are not able to arrive at any certain Conclusion of Science and Knowledge by such Disputations; and they find their Lives to be at an End before their Questions are so. When therefore all Things appear to be uncertain as to these Points, we must have our Recourse to the *True Prophet*, whom God the Father would have beloved by all: And in order thereto he would entirely put an End to these Inventions of Men, wherein there was no certain Knowledge to be found, that so
he

he might be the more enquir'd for, and might open that way of Truth to Men, which those others had shut from them. For his Sake God did also make the World; and by him the World is settled: Whence it is that he is every-where present to those that seek him after a Pure, and Holy, and Faithful Manner.

See this Matter further Prosecuted in my Chronology, p. 3—7.

Now from all this Evidence, and much more that might be alledg'd, it is apparent that the Jewish *and* Christian *Religions always suppose that there must be a due Temper of Mind in the Enquirers, or else the Arguments for those Religions will not have their due Effect. That the Course of God's Providence designs hereby to distinguish between the well-dispos'd, the Meek, the Humble, and the Pious; which are those whom the common Light of Nature declares may expect the Divine Blessing on their Studies of this sort; and the Ill-dispos'd, the Obstinate, the Proud, and the Impious; which are those whom the same common Light of Nature assures us may expect the Divine Malediction on the same: And that 'tis not for want of convincing and satisfactory Evidence in the Business of Revelation, but because many Men come with Perverse, Sceptical,*

a 3 *and*

and wicked Dispositions, that they fail of Satisfaction therein. Accordingly, I think it is true in common Observation, That the Virtuous and the Religious, I mean those that are such according to Natural Conscience, do rarely, if ever, fail on their Enquiries to Embrace and Acquiesce in both the Jewish *and* Christian *Revelations; and that the Debauch'd and Prophane do as seldom fail on their Enquiries to Reject and Ridicule them. Which different Success of the same Examination, agrees exactly with the whole Tenor of the Scriptures; and is the very same which must be true, in case those Scriptures be true also; and is, by Consequence, a considerable Confirmation of their real Verity and Inspiration. And certainly, he that considers his own Weakness and Dependance on God, and that all Truth and Evidence must come originally from him, will by Natural Judgment and Equity pronounce, that he who expects the Divine Blessing and Illumination, in Points of such vast Consequence, as those of Revelation most certainly are, ought above all Things to purify his Will, and rectify his Conduct in such Points as all the World knows to be the Will of God; and to Address himself to the Divine Majesty with due Fervency and Seriousness, for his Aid and Assistance, before he can justly promise himself Success in so great and momentous an Undertaking.*

But

PREFACE. xxiij

But then, as to the Second Enquiry, *or the* Degree *of* Evidence *that ought to be expected in Religious Matters, it seems to me very necessary to say somewhat upon this Subject also, before we come to our main Design. For as on the one Side it is a great Error in all Cases to expect such Evidence as the Nature of the Subject renders impossible; so is it as weak on the other Side, to lay the Stress of important Truths on such Evidence, as is in its own Nature Unsatisfactory and Precarious; or to assert with great Assurance what can no way be Proved, even by that sort of Evidence which is proper for the Subject in Debate. An Instance of the first Sort we have in* Autolicus, *an Heathen, in his Debates with* Theophilus *of* Antioch; *who appears weakly to have insisted upon* Seeing *the God of the Christians, e'er he would believe his Existence: While one of the known Attributes of that God is, that he is* Invisible. *And almost equally preposterous would any Philosophick Sceptick now be, who should require the* Sight *of the Air in which we Breathe, before he would believe that there was such an Element at all. Whereas it is clear, that the Air may be demonstrated to be sufficiently sensible and real, by a Thousand Experiments; while yet none of those Experiments can render it* Visible *to us: Just as the Existence of a Supreme Being may be*

Theoph. ad Autolyc, L. I. in initio.

a 4 *demon-*

demonstrated by innumerable Arguments, although none of those Arguments imply even the Possibility of his being properly Seen by any of his Creatures. But then, *that we may keep a Mean here, and may neither on one Side, expect in our Religious Enquiries,* overbearing, *or strictly* Mathematick Evidence, *such as is impossible to be deny'd or doubted of by any ; which would render the constant Design of Providence, already stated, entirely ineffectual, and* force *both Good and Bad to be Believers, without all Regard to their Qualifications and Temper of Mind: Nor on the other Side, may depend on such weak and precarious Arguments, as are not really sufficient nor satisfactory to even Fair, Honest, and Impartial Men : I intend here to Consider, what that Degree of Evidence is, which ought to be here insisted on; without which we are not, and with which we are obliged to acquiesce in Divine Matters.* Now *this Degree of Evidence I take to be that, and no other, which upright Judges are determin'd by in all the important Affairs of Estate and Life that come before them: And according to which, they ever aim to give Sentence in their Courts of Judicature. I chuse to Instance in this Judicial Evidence, and these Judicial Determinations especially, because the Persons concern'd in such Matters are, by long Use, and the Nature of their Employ-*

PREFACE.

Employment, generally speaking, the best and most sagacious Discoverers of Truth, and those that judge the most unbiass'dly and fairly, concerning sufficient or insufficient Evidence of all others. Such upright Judges then, never expect strictly Undeniable, or Mathematick Evidence; which they know is, in Human Affairs, absolutely impossible to be had: They don't require that the Witnesses they Examine, should be Infallible, or Impeccable, which they are sensible would be alike Wild and Ridiculous: Yet do they expect full, sufficient, or convincing Evidence; and such as is plainly Superior to what is alledged on the other Side: And they require that the Witnesses they believe, be, so far as they are able to discover, of a good Character, Upright and Faithful. Nor do they think it too much Trouble to use their utmost Skill and Sagacity in discovering where the Truth lies; how far the Witnesses agree with, or contradict each other; and which way the several Circumstances may be best compar'd, so as to find out any Forgery, or detect any Knavery which may be suspected in any Branches of the Evidence before them. They do not themselves pretend to judge of the Reality or Obligation of any Ancient Laws, or Acts of Parliament, from their own meer Guesses or Inclinations, but from the Authentickness of the Records which contain them; and though they

they are not able always to see the Reason, or Occasion, or Wisdom of such Laws, or Acts of Parliament; yet do they, upon full External Evidence that they are Genuine, allow and execute the same: As considering themselves to be not Legislators, but Judges: And owning that Ancient Laws, and Ancient Facts, are to be known not by Guesses or Supposals, but by the Production of Ancient Records, and Original Evidence for their Reality. Nor in such their Procedure do they think themselves guilty in their Sentences, if at any Time afterwards they discover that they have been impos'd upon by false Witnesses, or forged Records; supposing, I mean, that they are conscious that they did their utmost to discover the Truth, and went exactly by the best Evidence that lay before them; as knowing they have done their Duty, and must in such a Case be Blameless before God and Man, notwithstanding the Mistake in the Sentences themselves. Now this is that Procedure which I would earnestly recommend to those that have a Mind to enquire to good Purpose into Reveal'd Religion. That after they have taken Care to purge themselves from all those Vices, which will make it their great Interest that Religion should be false; after they have resolv'd upon Honesty, Impartiality, and Modesty, which are Virtues by the Law of Nature; after they have devoutly implor'd the

Divine

PREFACE.

Divine Assistance and Blessing on this their important Undertaking; which is a Duty likewise they are obliged to by the same Law of Nature; that after all this Preparation, I say, they will set about the Enquiry it self, in the very same Manner that has been already describ'd, and that all our upright Judges proceed by in the Discovery of Truth. Let them spare for no Pains, but consult all the Originals, whenever they can come at them: And let them use all that Diligence, Sagacity, and Judgment, which they are Masters of, in order to see what real External Evidence there is for the Truth of the Facts on which the Jewish *and* Christian *Religions do depend. I here speak of the* Truth of Facts, *as the surest way to determine us in this Enquiry; because all the World, I think, owns that if those Facts be true, these Institutions of Religion must also be true, or be deriv'd from God; and that no particular Difficulties, as to the Reasons of several Laws, or the Conduct of Providence in several Cases, which those Institutions no where pretend to give us a full Account of, can be sufficient to set aside the convincing Evidence which the Truth of such Facts brings along with it. For Example: Those who are well satisfy'd of the Truth of the* Mosaick *History of the Ten miraculous Plagues with which the God of* Israel *smote the* Egyptians; *of*

the

the drowning of the Egyptians *in the* Red-Sea; *while the* Israelites *were miraculously conducted through the same; and of the amazing manner wherein the Decalogue was given by God to that People at Mount* Sinai; *will, for certain, believe that the* Jewish *Religion was in the main derived from God, though he should find several occasional Passages in the* Jewish *Sacred Books, which he could not Account for, and several ritual Laws given that Nation, which he could not guess at the Reasons why they were given them. And the Case is the very same as to the Miraculous Resurrection, and Glorious Ascension of our Blessed Saviour,* Jesus Christ, *with Regard to the* New Testament. *On which Account I reckon that the Truth of such Facts is to be principally enquired into, when we have a mind to satisfy our selves in the Verity of the* Jewish *and* Christian *Religions. And if it be alledg'd that some of these Facts are too remote to afford us any certain Means of Discovery at this Distance of Time; I Answer, That then we are to select such of those Facts as we can examine, and to search into the Acknowledgment or Denial of those that are Ancienter, in the oldest Testimonies now Extant; into the Effects and Consequences, and standing Memorials of such Facts in After-Ages, and how far they were real, and allow'd to be so; and in short, we are to deter-*

determine concerning them, by the beft Evidence we can now have; and not let a bare Sufpicion, or a Wifh that Things had been otherwife, overbalance our real Evidence of Facts in any Cafe whatfoever. I do not mean that our Enquirer is to have no Regard to Internal Characters, *or the* Contents *of the* Jewifh *and* Chriftian *Revelations; or that he is not to examine into that alfo in the General, before he admits even the Proof from Miracles themfelves; becaufe what pretended Miracles foever are wrought, for the Support of Idolatry, or Wickednefs; for the Eftablifhment of Notions contrary to the Divine Attributes, or of an Immortal, or Prophane, or Cruel Religion, though they may prove fuch a Religion to be Supernatural, yet will they only prove that it comes from wicked Dæmons, or Evil Spirits, and not from a God of Purity and Holinefs, and fo will by no means prove it Divine, or worthy of our Reception. But then, it is, for the main, fo well known, that the* Jewifh *and* Chriftian *Inftitutions do agree to the Divine Attributes, and do tend to Purity, Holinefs, Juftice, and Charity; and are oppofite to all Immorality, Prophanenefs, and Idolatry, that I think there will not need much Examination in fo clear a Cafe; and that, by Confequence, our main Enquiry is to be as to the Truth of the Facts thereto relating. And in*

this

this Case, I fear not to Invite all our Scepticks and Unbelievers, to use their greatest Nicety, their entire Skill, their shrewdest Abilities, and their utmost Sagacity in this Enquiry; being well assur'd from my own Observations in this Matter, That the proper Result of such an exact Historical Enquiry will be as plainly and evidently on the Side of Reveal'd, *as I have demonstrated in this Treatise, that Philosophy and Mathematicks are on the Side of both* Natural *and* Reveal'd Religion. *And now having Premis'd this, I come to my main Design; to shew what is properly the* Religion *of a genuine and considering* Astronomer; *or what are properly the* Astronomical Principles *of Natural and Reveal'd Religion.*

Mr.

Mr. *Milton's* HYMN

TO THE

CREATOR.

THese are thy glorious works, Parent of good,
　Almighty, thine this univerſal Frame,
Thus wondrous fair; thy ſelf how wondrous then!
Unſpeakable, who ſit'ſt above theſe Heavens
To us inviſible, or dimly ſeen
In theſe thy loweſt Works; yet theſe declare
Thy Goodneſs beyond Thought, and Power Divine:
Speak ye who beſt can tell, ye Sons of Light,
Angels, for ye behold him, and with Songs
And choral Symphonies, Day without Night,
Circle his Throne rejoycing: ye in Heav'n,
On Earth joyn all ye Creatures to extoll
Him firſt, Him laſt, Him midſt, and without End.
Faireſt of Stars, laſt in the train of Night,
If better thou belong not to the Dawn,
Sure Pledge of Day, that crown'ſt the ſmiling Morn
With thy bright Circlet, praiſe him in thy Sphere
While Day ariſes, that ſweet Hour of Prime.
Thou Sun, of this great World both Eye and Soul,
Acknowledge him thy Greater, ſound his Praiſe
In thy eternal Courſe, both when thou climb'ſt,
And when high Noon haſt gain'd, and when thou fall'ſt.
Moon, that now meet'ſt the orient Sun, now fly'ſt,
With the fixt Stars, fixt in their Orb that flies,

　　　　　　　　　　　　　　　　And

And ye Five other wandring Fires that move.
In myſtic Dance, not without Song, reſound
His Praiſe, who out of Darkneſs call'd up Light.
Air, and ye Elements, the eldeſt Birth
Of Nature's Womb, that in quaternion run
Perpetual Circle, multiform; and mix
And nouriſh all Things, let your ceaſeleſs Change
Vary to our great Maker ſtill new Praiſe.
Ye Miſts and Exhalations that now riſe
From Hill or ſteaming Lake, duſky or grey,
Till the Sun paint your fleecy Skirts with Gold,
In Honour to the World's great Author riſe:
Whether to deck with Clouds the uncolour'd Sky,
Or wet the thirſty Earth with falling Showers,
Riſing or falling ſtill advance his Praiſe.
His Praiſe ye Winds that from four Quarters blow,
Breath ſoft or loud; and wave your tops, ye Pines,
With every Plant, in ſign of Worſhip wave.
Fountains, and ye, that warble, as ye flow,
Melodious murmurs, warbling tune his Praiſe.
Joyn Voices all ye living Souls, ye Birds,
That ſinging up to Heaven's high Gate aſcend,
Bear on your Wings and in your Notes his Praiſe;
Ye that in Waters glide, and ye that walk
The Earth, and ſtately tread, or lowly creep;
Witneſs if I be ſilent, Morn or Even,
To Hill, or Valley, Fountain, or freſh Shade
Made Vocal by my Song, and taught his Praiſe.
Hail univerſal Lord! be bounteous ſtill
To give us only good; and if the Night
Have gathered ought of evil or conceal'd,
Diſperſe it, as now Light diſpels the Dark.

Paradiſe Loſt, Lib. V.

Aſtrono-

Astronomical Principles
OF
RELIGION,
Natural, and Reveal'd.

Part I.

Lemmata:

Or, *The known Laws of* Matter *and* Motion, *preparatory to the ensuing Treatise.*

(*Taken out of the* Author's *Mathematical Philosophy, where they are all demonstrated.*)

1. EVERY Body perseveres in its own present State, whether it be that of Rest, or uniform direct Motion; unless it be compelled by some Force impress'd, to change that State.

(2.) All

Astronomical Principles

(2.) All Motion is of it self Rectilinear.

(3.) All revolving Bodies endeavour to recede from the Center of their Motion; and by how much the Motion is the swifter, this Endeavour is the greater.

(4.) The Mutation of Motion is proportional to the moving Force impress'd; and is according to the Direction of that Line along which that Force is impress'd.

(5.) Re-action is always contrary and equal to Action. That is, the Actions of Two Bodies acting upon each other, whether they be Impulses or Attractions, are always in opposite Directions, and are also equal.

(6.) If of two equal Bodies, void of Elasticity, one of them which is in Motion meets the other at rest, upon the meeting they will both proceed forwards together, to the same part, with half the Velocity of the Body which was moved.

(7.) If two equal Bodies, void of Elasticity, do directly meet each other with the same Velocity, they upon the Collision will both of them rest.

(8.) If two unequal Bodies, destitute of Elasticity, meet one another with such Velocities, that by how much the greater exceeds the other in Magnitude, by so much it is exceeded by the lesser in Swiftness, so that the Velocities are reciprocal to the Bodies; they will both rest after that meeting.

(9.) If a moving Body strike another at rest, (but both void of Elasticity) how unequal soever they be in Bulk and Quantity of Matter, they will both move after the shock with the same Velocity towards the same Parts, as in the

Sixth

of RELIGION.

Sixth Law: And the common Velocity will be so much less than the first, as both the Bodies together are greater than the Body first moved.

(10.) If two unequal Bodies, void of Elasticity, which are moved with equal Velocity to opposite Parts, hit against one another, the Quantity of Motion in both, taken together after the Collision, will be the *Difference* only of the former Motions.

(11.) If two equal Bodies, void of Elasticity, be mov'd with unequal Velocity towards the same Part, upon their Collision there will remain the same Quantity or *Sum* of their Motions; but the common Velocity will be only the half of both the former Velocities put together.

(12.) If of two unequal Bodies, void of Elasticity, the Greater overtakes the Lesser, the common Velocity, after the Shock, will be greater than half the Sum of the former Velocities. And on the contrary, it will be less when the lesser Body overtakes the greater.

(13.) If a Body perfectly Elastic dasheth upon another Body of the same sort which is Quiescent and Equal; after the Collision the Motion will be wholly transferr'd into that which was quiescent before, and with the same Celerity; but the Body which was mov'd before, will now rest.

(14.) If two Bodies perfectly Elastic, which are equal, but mov'd with an unequal Celerity, dash one upon another, they, whether they were before carried to the same part, or to the contrary, will, after the Contact, be mov'd each with that Celerity which the other had before.

B 2 (15.) Any

(15.) Any Body, how great foever, may be moved by any Body, how fmall foever, coming with any Velocity whatfoever.

(16.) When two Bodies, perfectly Elaftical, are dafh'd one upon the other, they depart from one another with the fame Celerity wherewith they approach'd one to the other; that is, not with the fame *abfolute*, but *relative* Celerity.

(17.) If two Bodies perfectly Elaftical, do each return to the Impulfe with the fame Celerity wherewith they rebounded from it; they will each of them, after the Second Impulfe, acquire the fame Celerity as they had before the firft Meeting.

(18.) If two Bodies meet one another, whether they be Elaftic or not Elaftic, there doth not always remain the fame Quantity of Motion as was before, but it may be greater or lefs.

(19.) If a Body perfectly Elaftical, which is greater, hits upon a leffer one which is quiefcent, it will give a Velocity to it lefs than the double of its own.

(20.) If two Bodies perfectly Elaftic, the Celerities whereof are in reciprocal Proportion to their Magnitudes, meet one another directly and oppofitely, they will both rebound with the fame Celerity with which they came to each other.

(21.) The Celerity which a greater Body perfectly Elaftic, gives to a leffer perfectly quiefcent, which is alfo perfectly Elaftic, hath that Proportion to that Velocity, which the leffer moved with the like Celerity gives to the greater when quiefcent, which the Magnitude of the greater hath to the Magnitude of the lefs.

(22.) Every

of RELIGION.

(22.) Every Body will in the same Time describe the Diagonal of a Parallelogram with Forces conjunct, that it would do the Sides with those Forces separate.

(23.) All compound Forces and Motion whatever may be reduc'd into innumerable other direct Forces and Motions; and on the contrary, all direct Forces, and rectilinear Motions, may be suppos'd to be compounded of innumerable oblique Motions and Forces.

(24.) The Quantity of Motion which is collected, by taking the Sum of the Motions to the same Part, and the Difference of those to the contrary Parts, is not chang'd by the Actions of Bodies one upon another.

(25.) The common Center of Gravity of a System of Bodies doth not change its State either of Motion or Rest, from the Actions of the Bodies amongst themselves, (whether they be Attractions or Impulses;) and therefore the common Center of Gravity of all Bodies acting upon one another (Actions and Impediments, whether External or otherwise arising, being excluded) doth either rest, or is mov'd uniformly straight forwards.

(26.) The Motions of two Bodies included in a given Space, and partaking of the Motion thereof, are the same amongst themselves, whether that Space resteth, or the same is mov'd uniformly straight forward, without a Circular Motion.

(27.) If Bodies be mov'd in any wise amongst themselves, and be pressed with equal accelerative Forces according to parallel Lines, they will all continue to be mov'd in the same manner amongst themselves, as if they were not pressed with those Forces.

PROP

Propositions.

III. The Velocities of a Body accelerated by any uniform urging Force whatever, are betwixt themselves, as the Times are wherein that uniform Force is impress'd; that is, in Double the Time Double, in Triple the Time Triple, and in Four Times the Time Quadruple.

IV. The Lines which Bodies by any urging uniform Force do describe, are in the duplicate proportion of the Times, *i. e.* if the Times be Seconds, One, Two, Three, Four, Five, *&c.* the whole Lines describ'd will be amongst themselves, as One, Four, Nine, Sixteen, Twenty-five, *&c.* which are the Squares of the former.

VII. In a Cycloid inverted, whose Axis is erected perpendicular, the Times of the Descent wherein a Body let down from any Point whatever in it, comes to the lowest Point, are always equal betwixt themselves.

VIII. All Projectiles, not perpendicular to the Horizon, describe Parabola's, so far as they are not hindred by the resistance of the Air.

IX. If two Bodies do in equal Times run over Two whole unequal Circumferences, with an equable Motion, the centripetal Force in the greater Circumference will be to that which is in the less, as the Circumferences are one to another directly; or, which is the same, as their Diameters, or Radii.

X. If two Bodies revolve in the same, or equal Circles with unequal Celerities, but both with

an

an equable Motion, the centripetal Force of the Swifter will be to that of the Slower, in the Proportion of the Celerities duplicated; or as the Squares of the Arches described together.

XI. If two Bodies revolve in unequal Circles with equal Velocity, their centripetal Forces will be in the reciprocal Proportion of their Circumference or Diameters; so that in the lesser Circumference there will be the greater centripetal Force, and in the greater the lesser.

XII. If two Bodies be mov'd in unequal Circles, with an unequal Velocity, in the sub-duplicate Proportion of the Circumferences, Diameters, or Radii, the centripetal Forces will be equal every where, and neither increas'd in the Access nor Recess.

XIII. If two Bodies be mov'd in unequal Circles, with an unequal Velocity, in the sub-duplicate Proportion of the Circumferences, Diameters, or Radii, reciprocally; so that in the greater Circle the Velocity be the lesser, and in the lesser Circle the greater, and this in the said sub-duplicate reciprocal Proportion, the centripetal Force will be reciprocally as the Squares of the Radii or Distances.

XIV. If two Bodies revolve in unequal Circles with an unequal Celerity; so that by how much greater the Radius, Diameter or Circumference is, so much the less the Velocity is; and by how much the less the Radius is, so much the greater is the Velocity, and this in the Reciprocal Proportion of the Radii, the Centri-petal Forces will be as the Cubes of the Radii reciprocally.

XV. The Area's, which revolving Bodies do describe by Radii drawn unto the unmovable

Center of Force acting upon them, do both lie in immoveable Planes, and are proportional to the Times; and so in any given Time are everywhere equal; the Velocity of Motion in the lesser Distance, and the Slowness thereof in the greater so tempering the Description of the Area's, that from those various Distances no Difference of the Spaces run over in the given Time doth ever arise.

XVI. Every Body which is mov'd in a Curve Line, and doth by a Radius drawn to some Point, either immoveable, or going forwards uniformly with a Rectilineal Motion, describe Areas about that Point proportional to the Times; is urged or impress'd by a Centripetal Force tending to the same Point.

XIX. If a Body be mov'd in an Ellipsis about the Center of the same, the Centripetal Force will be directly as the Distance of the Body from the same Center.

XX. If a Body be mov'd in a Spiral Line, which cuts all the Radii in the same Angle, the Centripetal Force will be reciprocally as the Cube of the Distance from the Center of the Spiral.

XXI. If a Body be mov'd in an Ellipsis or Parabola, or Hyperbola, about its Focus, the Centripetal Force will be every where in the duplicate Proportion of the Distance from the same Focus reciprocally.

XXII. The Velocity of a Body moving in a Parabola about a Body placed in the Focus, the Force whereof is in the reciprocal duplicate Proportion of the Distances, is every where to the Velocity of a Body revolving in a Circle in the same time, in the subduplicate Proportion of the

the Number, Two to Unity; or as the Diagonal of a Square to its Side; that is, as 10 to 7 nearly.

XXVII. Two Bodies attracting one another, describe like Figures both about the Common Center of Gravity, and about one another; that is, whilst they really describe like Figures about the Common Center of Gravity, the Eye being placed in either of the two, and not perceiving its own Motion, or that of the Center of Gravity, a Figure like to the same will thereby seem to it to be describ'd.

XXXI. If a primary Planet revolving about the Sun carry a Moon along with it, this will be so mov'd about the Primary, that it will perpetually be accelerated from the Quadrature with the Sun, unto the Conjunction or Opposition next following; but from the Conjunction to the Quadrature, it will be retarded; and consequently will be carried more swiftly about the Conjunction and Opposition, but more slowly about the Quadratures.

XXXVIII. The absolute Force of the Sun in the disturbing the Secondary Planets, and the Effects thereof, in divers Distances from the Sun, is in the triplicate Proportion of those Distances inversly.

XLI. If a Fluid be contain'd in a Channel form'd in the Surface of any Planet, Primary or Secondary, and be uniformly revolv'd together with the Planet with a diurnal periodic Motion; each Part of this Fluid will be accelerated and retarded by turns; in its Conjunction and Opposition, or at Noon-day and Midnight, it will be swifter; in the Quadratures, or at the 6th Hour Evening and Morning, it will be slower than

than the contiguous Surface of the Globe; and thus there will be a flux and reflux in the Channel, by turns perpetually.

XLII. If a Solid Ring be put about a Globe perfectly spherical, at the Equator of the same, and stick to it; there will indeed be no Motion of Flux and Reflux, but the vibrating Motion of Inclination, and the Precession of the Nodes, will remain. Let the Globe have the same Axis with the Ring, and compleat its Revolution in the same time; and with its Surface touch the Ring inwardly, and cleave to it; by its participating of the Motion thereof, the whole Frame will vibrate to and fro, and the Nodes will go back.

XLIV. If towards each equal Points of a Spherical Physical Surface of equal Thickness every where, but which Thickness is so small that it is not to be regarded, there be a Tendency of equal Centripetal Forces decreasing in the duplicate Proportion of the Distances from the same Points; any Corpuscle placed any where within this Surface, will not be attracted unto any Part by the said Force; but will either rest, or continue that Motion which is begun without any Disturbance, and in the same manner as if it were acted upon with no Force at all from that Surface: And the case is the same in any Spherical concave Space within a solid Sphere, about its Center.

XLVII. If unto each Point of some given Sphere, which is Homogeneous, or of equal Density every where, there be a Tendency of equal Centripetal Forces decreasing in the duplicate Proportion of the Distances from the Points; a Corpuscle placed within the Sphere, is attracted

with

with a Force proportional to its Distance from the Center thereof.

LIX. If the Density of a Fluid, compos'd of Particles which do flee from each other, be as the Compression; so that if the pressing Force be two, or four, or eightfold, the Density thence arising is so likewise; the Centrifugal Force of the Particles is reciprocally proportional to the Distances from the Center: And, *vice versâ*, where the said Force is reciprocally proportional to the Distances from their Centers, the Particles which flee from each other compose an elastic Fluid, the Density whereof is proportional to the Compression.

LX. The Quantity of Matter in all Bodies, is exactly proportional to their Weight.

LXII. Bodies mov'd with an unequal Velocity in a very Subtle Fluid, are resisted by the Fluid in the Duplicate Proportion of their Velocity.

LXIV. As the Resistance of Fluids in divers Velocities is in the duplicate Proportion of the Velocity; so in divers Densities the Velocity being given, it is in the direct Proportion of the Density it self; but the Density and Velocity being given, in the duplicate Proportion of the Diameters; and consequently the Resistance in general is in a Proportion compounded of the duplicate Proportion of the Velocity, and the duplicate Proportion of Diameters, and the simple Proportion of the Density of the Medium directly.

LXVII. If a solid Cylinder, infinitely long, be revolv'd in an uniform and infinite Fluid about its own Axis, the Position whereof is given, and the Fluid be mov'd round by the Impulse

pulse of this Cylinder only; and every Part of the Fluid perseveres uniformly in its Motion; the periodic Times of the Fluid will be as their Distances from the Axis of the Cylinder directly; and the Velocities will be every where equal.

LXVIII. If a solid Sphere, in an uniform and infinite Fluid, be revolv'd uniformly about its own Axis, the Position whereof is given; and by the Impulse of this alone the Fluid be turned round, and every part of the Fluid perseveres uniformly in its Motion; the periodic Times of the Parts of the Fluid will be as the Squares of the Distances from the Center of the Sphere.

LXIX. The Velocities of all the Planets, whether Primary or Secondary, about their Central Bodies, by being in the reciprocal subduplicate Proportion of the Distances from their Centers, do wholly overthrow the *Cartesian* Hypothesis of Vortices.

LXX. The Six Primary Planets, each with its own Satellites, where they have any, encompass the Sun with their Orbs, and revolve about it.

LXXI. The periodic Times of the six Primary Planets, are in the sesqui-alteral Proportion of their mean Distances from the Sun.

LXXII. The six Primary Planets do always, by Rays drawn to the Sun, describe equal Areas in equal Times, and in general Areas proportional to the Times.

LXXIII. The Moon, by Rays drawn to the Center of the Earth, describes in equal Times Areas almost equal; and in general, Areas almost proportional to the Times.

LXXIV. The

of RELIGION.

LXXIV. The Satellites of *Jupiter* do, by Rays drawn to the Center of *Jupiter*, describe Areas proportional to the Times: And their periodic Times are in the sesqui-alteral Proportion of their Distances from the Center of their Primary.

LXXV. The Satellites of *Saturn* do, by Rays drawn to the Center of *Saturn*, describe Areas proportional to the Times: And their periodic Times are in the sesqui-alteral Proportion of their Distances from the Center of their Primary.

LXXVI. The Force whereby the Primary Planets are perpetually drawn back from right Lines, and retain'd in their Orbs, does respect the Sun; and is as the Squares of the Distances from the Center of the Sun reciprocally.

LXXVII. The Force wherewith the Satellites of *Jupiter* and *Saturn* are perpetually drawn back from right Lines, and retain'd in their Orbs, respect the Centers of *Jupiter* and *Saturn* respectively; and is as the Squares of the Distances from those Centers reciprocally.

LXXVIII. The Force wherewith the Moon is perpetually drawn back from a Rectilinear Motion, and retain'd in its Orb, respects the Center of the Earth; and is as the Squares of the several Distances from the same Center reciprocally.

PART

PART II.

A particular Account of the System of the Universe.

THE *Sun*, that immense and amazing Globe of Fire, the Fountain of all the Light and Heat of the whole Planetary and Cometary World, is in Diameter 763 000 Miles, in Surface it contains 1,813.200,000.000 Square Miles, and in Solidity 23.000,000.000,000 000 Cubical Ones, in Magnitude 900.000, and in Quantity of Matter 230.000 Times as great as the Earth, tho' only a Quarter so Dense; and all Bodies weigh 24 Times as much on its Surface, as on the Surface of the Earth. It is situate near the Center of Gravity of the whole System, and revolves in about 25 Days and a half round its own Axis. It has frequently Spots, and sometimes brighter Parts seen upon its Surface, of vast Dimensions; as if they were great burning Vulcanes, sometimes clouded with Smoke, and sometimes clear. Its Heat, on its own Surface

of RELIGION. 15

is above 11000 Times as Intense as that on the Earth. All the Planets and Comets gravitate to the *Sun* in a duplicate reciprocal Proportion of their Distances from it, and are thereby retained in their several Orbits. Their Periodical Times are in a sesquiplicate or sesquialteral proportion to their Distances; that is, the Triplicate or Cubes of the Distances, are as the Duplicate or Squares of the periodical Times; and that to the greatest Exactness possible; which equally obtains in the secondary Planets, with regard to their primary Ones also; and is the fundamental Law of the entire System.

Mercury is the nearest to the *Sun* of all the known Planets. Its utmost Elongation from it, to an Eye on the Earth is but 28 Degrees, so that it is but rarely seen by us. This Planet is in Diameter 4.248 Miles, in Surface it contains 55,000.000 of square Miles; and in Solidity 39.000,000.000 of Cubical Ones, & is 32,000.000 Miles distant from the *Sun*, and describes a very Eccentrical Ellipsis about it in less than 3 Months, or in 88 Days. The Eccentricity of its Orbit is $\frac{210}{1000}$ of its mean Distance from the *Sun*; and by its Position must appear thro' a Telescope with Phases like those of the Moon. No secondary Planets have yet been observed about it, nor any Diurnal Rotation. It enjoys above 6 Times as much Light and Heat from the Sun, as doth the Earth; and it appears very rarely like a Spot in the Disk of the Sun, in its Retrograde Conjunctions, when it passes between the Sun and Earth.

Venus is somewhat higher in the System, and so has its utmost Elongation 45 Degrees. It is a larger Planet than *Mercury*, and comes sometimes

times much nearer to us. It is our Morning and Evening Star by turns, and is the Brightest of the Heavenly Bodies to us, next the Moon, as casting a visible Shadow in the Dark, and sometimes appearing in the very Day-time also. It is in Diameter 7.900 Miles; in Surface it contains 2.000,000.000 of Square Miles; and in Solid it is 264.000,000.000 of Cubical Ones; and is distant 59,000.000 from the Sun, and describes its Ellipsis about it in 7 Months and a half, or 225 Days. The Eccentricity of its Orbit is but $\frac{7}{1000}$ of its mean Distance from the Sun. It most plainly thro' the Tellescope appears with Phases, and Horned like the Moon. No secondary Planets have yet been discovered about it, yet has it a Diurnal Revolution on its own Axis in 23 Hours. It receives almost double the Light and Heat from the Sun which the Earth does; and appears very rarely as a Spot in the Sun also.

The *Earth* is the next Planet to *Venus*, and has the Moon for its secondary Planet; the common Center of whose Gravity describes an Ellipsis about the Sun in one Year, or 365 Days and a Quarter, nearly; the Eccentricity of its Orbit is $\frac{17}{1000}$ of its mean Distance from the Sun, in Surface it contains 200,000.000 of Square Miles, and in Solidity 266.000,000.000 of Cubical Ones, and is in Diameter 7.970 Miles. and is distant from the Sun 81,000.000 Miles. This Annual Motion is perform'd in the Ecliptick, and is directed, as is that of all the Planets, primary and secondary, from *West* to *East*, or according to the Order of the Signs, and therefore causes the Sun to have an apparent Annual Motion the same way, and in the

same

of R ELIGION. 17

same Plain; but as still in the opposite Point of the Ecliptick. It has also a Diurnal Rotation upon its own Axis from *West* to *East* in 24 Hours, and so occasions an apparent Motion of all the Heavenly Bodies from *East* to *West* in the same time. The Axis of the Diurnal Motion is 23 Degrees and one half, oblique to that of the Ecliptick, which occasions the Varieties of *Spring*, *Summer*, *Autumn*, and *Winter*. It is in Figure an Oblate Spheroid, as having the Diameter of its Equator about 62 Miles longer than its Axis, on Account of the Elevation of the Equatorial and Depression of the Polar Regions, occasion'd by the centrifugal Force of the Diurnal Motion.

Mars is still higher in the System, and looks more red and fiery than the rest of the Planets. It takes a larger Circuit than the Earth, and so comes to its Conjunction, Quadratures and Opposition; and in some Degree imitates the Phases of the Moon, being sometimes only Gibbous, tho' it cannot be at all Horned like the other. This Planet describes its Ellipsis about the Sun in less than 2 Years, or in 687 Days. The Eccentricity of its Orbit is $\frac{91}{1000}$ of its mean Distance from the Sun; it is in Diameter 4444 Miles; in Surface it contains 60,000,000 of Square Miles; and in Solidity 44,000,000,000 of Cubical Ones; and is distant from the Sun 123,000,000 Miles; it has no secondary Planet that can be seen, but revolves about its own Axis in 24 Hours and 40 Minutes. The Quantity of Light and Heat it enjoys from the Sun is between one half and one third of what the Earth receives from it. It also appears to us upon the Earth to be sometimes Direct, some-

C

times Stationary, and sometimes Retrograde in its Course, as all the superior Planets most remarkably do; which Appearances, their higher Situation, and slower Angular Motion, or longer Periods, must necessarily produce, without the least Alteration of their real progressive Motion about the Sun all the while.

Jupiter, the largest of all the Planets, is much higher in the System, and has four Satellites or Moons revolving about it; and all by their common Center of Gravity describe a very great Ellipsis about the Sun. The Eccentricity of its Orbit is $\frac{48}{1000}$ of its mean Distance from the Sun. It comes to its Conjunction, Quadratures, and Opposition, as well as *Mars*; but at so great a Distance can never appear other than full, or nearly so. It is in Diameter 81.000 Miles; in Surface it contains 20.000,000.000 of Square Miles, and in Solidity 280.000 000, 000.000 Cubical Ones; and revolves about the Sun in Eleven Years and Ten Months, or 4332 Days and a Half, at the Middle Distance of 424,000.000 Miles. It revolves about its own Axis in 9 Hours, and 56 Minutes, which makes its Figure that of an Oblate Spheroid, having the Diameters of its Equator considerably longer than its Axis. The Quantity of Light and Heat it receives from the Sun is but one Twenty-seventh Part so great as ours on the Earth. Its Quantity of Matter is about 220 Times so great as that of the Earth. Its Density is about one fifth Part of the Earth's, and so the Weight of all Bodies on its Surface is about double to that with us. It is also Direct, Stationary and Retrograde as *Mars*, but not in so great a Degree. It has Belts, like Clouds, lying

of RELIGION.

lving somewhat regularly along the Equatorial Parts, but subject to many Changes and Variations.

Saturn, the highest and most remote of all the known Planets, has five Satellites or Moons, and a vast but thin Ring encompassing his Body, as an Horizon does a Globe; all which, or rather the common Center of their Gravity, describes an Ellipsis about the Sun. The Eccentricity of its Orbit is $\frac{65}{1000}$ of its mean Distance from the Sun. It comes to its Conjunction, Quadratures and Opposition, as well as the two former, but with no visible Decrease of its Light at its Quadratures, which is scarce to be expected at so great a Distance. It revolves about the Sun in about 29 Years and a half, or about 10.760 Days. It is in Diameter 68.000 Miles; in Surface it contains 14.000,000.000 of Square Miles, and in Solidity 160,000 000, 000.000 of Cubical Ones, at the mean Distance of 777,000.000 Miles from the Sun. Its Quantity of Matter is about 94 Times as great as that of the Earth, tho' its Density be only between a sixth and a seventh Part so great as that of the Earth's; and the Weight of Bodies on its Surface is to that on the Surface of the Earth as about five to four. It is not yet certainly known to revolve about its own Axis, tho' its Ring is said to do so. The Light and Heat communicated to it by the Sun are not quite the Ninetieth Part so great, as those bestow'd on the Earth. It is also in some measure Direct, Stationary and Retrograde, as well as the two former Planets, tho' still in a less Degree.

Of the *Satellites*, or *secondary Planets*; the most eminent as to us is the *Moon*. It describes an Ellipsis about our Earth, (or rather both Earth and Moon describe their own similar Ellipses about the common Center of their Gravity, as is the Case in all such Systems) in a periodical Month of 27 Day 57 Hours 43 Minutes. The Mean Eccentricity of its Orbit is $\frac{55}{1000}$ of its mean Distance from the Earth. It makes a Lunation or Synodical Month in 29 Days 12 Hours 44 Minutes. At a Mean it is distant from the Earth about 240,000 Miles, tho' with considerable Difference on Account of its great Eccentricity. The Moon's Diameter is 2175 Miles; in Surface it contains 14,000.000 of Square Miles; and in Solidity 5.000,000,000. of Cubical Ones. It has between the 39th and 40th part of the Quantity of Matter of the Earth; its Density is to that of the Earth as about 5 to 4, whilst the Weight of all Bodies is but about a third Part so great on its Surface, as on that of the Earth. It revolves from *West* to *East* upon its own Axis, exactly in a periodical Month, and thereby turns in general the same Face towards the Earth continually; yet does the Inequality of its Motion, and the Obliquity of its Axis, occasion some unequal *Librations* here also. It has very high Mountains, and very deep regular Valleys, and has lately had an Atmosphere discover'd about it; nor does it seem unlike the Earth as to Sea and Land. It has all variety of Phases, according to its various Position with respect to the Sun, or according as we on the Earth can see the whole, the half, or only some Part of its enlighten'd Hemisphere. Its own Day and Night are

are each half a synodical Month, or near 15 of our Days long. Its Orbit is inclin'd to that of the Ecliptick, at the least in an Angle of 5 Degrees; so it but sometimes passes just between the Sun and Earth at the new Moon, and but sometimes falls into the Earth's Conical Shadow at the Full. Yet when it is in, or near the Nodes, or Intersection of those Plains, it cannot avoid those Accidents; from the former of which the *Solar*, and from the latter of which the *Lunar Eclipses* are derived.

Jupiter's four Satellits, or secondary Planets, are visible with an ordinary Telescope, and sometimes pass like Spots on the Face of *Jupiter*, and sometimes enter into his Shadow; which to an Eye in *Jupiter* would cause Appearances just like our *Solar* and *Lunar Eclipses*. They revolve about him in Circles, or Ellipses very little Eccentrical; the Innermost at 130.000 Miles distance, in 1 Day 18 Hours and a half; the next at 364,000 Miles distance, in 3 Days 13 Hours and a quarter; the Third at 580,000 Miles distance in 7 Days $3\frac{1}{3}$ Hours; the Fourth at 1,000.000 Miles distance, in 16 Days $16\frac{2}{7}$ Hours. The Cubes of their Distances are also as the Squares of their periodical Times: and so they are kept in their Curvilinear Orbits by their Gravity towards *Jupiter's* Center, in a duplicate reciprocal Proportion from it; as is the Case of all the Planets, both Primary and Secondary about the Sun.

Saturn has *five Satellites* or secondary Planets. The Fourth in order from *Saturn* is the largest, and was discovered by the celebrated *Hugenius*; the Third and Fifth are visible in the next Degree; but the Knowledge of the two Innermost are wholly owing to *Cassini*'s extraordinary Glasses and Diligence. They all revolve in Orbits almost Circular, and are all in or very near the Plain of his Ring, which is inclined in an Angle of 31 Degrees to that of the Ecliptick. The innermost revolves about *Saturn* at 146,000 Miles distance, in 1 Day 21 Hours one third; the next at 187.000 Miles, in 2 Days, 17 Hours, and two thirds; the Third at 263,000 Miles, in 4 Days, 13 Hours, three Quarters; the Fourth or large one at 600,000 Miles in 15 Days 22 Hours two thirds; the last at 1,800.000 Miles, in 79 Days 22 Hours. Nor is it improbable, that the large Interval between the Fourth and Fifth may have a Sixth, which is yet to us invisible, as *Hugenius* conjectures. As to that strange and unparallel'd Phænomenon of *Saturn's Ring*, which is commonly visible through an ordinary Telescope; its Thickness may well be 500, or perhaps 1000 Miles, tho' it be at that Distance almost invisible; its Breadth is certainly about 21,000 Miles, and its distance from the Body of *Saturn* on every Side as much. It causes many different Appearances, not only to us on Earth, but much more to the Inhabitants of *Saturn*, if any such there are; all which *Hugenius* has describ'd in his System of that Planet, and others from him.

As to the *System* of the *Comets*, it appears now to be very considerable, and indeed they are the most numerous Bodies of the entire *Solar System*. They appear both by their Bigness and Motions to be a sort of Planets, revolving about the Sun in Ellipses, so very oblong, that their visible Parts seem in a manner Parabolical; but have such vast Atmospheres about them,

and

and Tails deriv'd from the same, especially after their *Perihelia*, and those subject to such Mutations, pass thro' so much Cold and Darkness near their *Aphelia*, and so much Light and Heat near their *Perihelia*, as imply them design'd for very different Purposes from the Planets; and indeed, as to their outward Parts, in their present State they are plainly uninhabitable. Yet by passing through the Planetary Regions in all Plains and Directions, they fully prove those Spaces to be destitute of Resistance or Solid Matter, and seem fit to cause vast Mutations in the Planets, particularly in bringing on them Deluges and Conflagrations, according as the Planets pass through the Atmosphere, in their Descent to, or Ascent from the Sun; and so seem capable of being the Instruments of Divine Vengeance upon the wicked Inhabitants of any of those Worlds; and of burning up, or perhaps, of purging the outward Regions of them in order to a Renovation. This, I mean, seems likely to be their use in the present State; tho' indeed they do withal seem at present *Chaos*'s or Worlds in Confusion, but capable of a Change to Orbits nearer Circular, and then of settling into a State of Order, and of becoming fit for Habitation like the Planets; but these Conjectures are to be left to farther Enquiry, when it pleases the Divine Providence to afford us more Light about them: However, in my *Solar System* I have described the Orbits of all the Comets that Dr. *Halley* has put into his Catalogue, and that in the Order of their Nearness to the Sun, at their *Perihelia*, and as they are in their proper Plains, without any reduction to the Ecliptick. They are in Number 21; for tho' he has 24 there set down,

down, (all which are accordingly numbered there) yet becaufe he fcarcely doubts that three of them are the fame Comet, and guefles that two more are alfo the fame, in both which Cafes I fully agree with him; the real Number will then be but 21. The former of thefe two (which alfo feems to have appeared before his Catalogue begins, *Anno Domini* 1456.) was feen in 1531, 1607, and 1682, whofe Period therefore is 75 or 76 Years, and whofe Return is to be expected in 1758. The latter of them appeared *Anno Dom.* 1532; and probably the fame again in 1661, whofe Period therefore being about 129 Years, it is to be expected again in 1789. The moft eminent of them all appear'd in the 44th Year before the Chriftian Æra; as alfo A. D. 531, or 532; and A. D. 1106; and laftly, A. D. 1680, 1681, when I faw it; and fo has made within the Limits of our prefent Hiftories, three periodical Revolutions, in about 575 Years apiece. The middle diftance of the former from the Sun's Center muft be 1458,000.000 of Miles, and its longer Axis twice fo long, and fo its *Aphelion* Diftance near four Times as great as the diftance of *Saturn*; and its greateft diftance to its leaft as about 60 to 1; and therefore its greateft Light and Heat to its leaft as about 3600 to 1. The middle Diftance of the fecond muft be about 2025,000.000 of Miles, and its longer Axis twice fo long, and fo its *Aphelion* diftance between 5 and 6 times as great as the diftance of *Saturn*; and its greateft Diftance to its leaft, as more than 100 to 1, and therefore its greateft Light and Heat to its leaft, as more than Ten Thoufand to One. The Middle Diftance of the laft muft be about

5.600,

5.600,000.000. Miles; and its longer Axis twice so long; and so its Aphelion Distance about 14 times as great as the distance of *Saturn*; and its greatest distance to its least as above 20.000 to 1. and so its greatest Light and Heat to its least as above 400,000.000 to 1.

As to the *Fixed Stars*, they are vastly remote from this our Planetary and Cometary System, but may perhaps every one be the Center of another such like System. Dr. *Hook* and Mr. *Flamsteed* think they have discover'd their Annual Parallax, and that it is about 45″, which will imply them to be about 700,000,000.000 of Miles distant from the Sun; or, according to an exact Calculation in the like Case, farther than a Bullet shot out of a Musket would go in 5000 Years. But of such vast and numberless Systems, if such they are, we know very little: Only so much we know of the Planetary and Cometary World, and of the Probability of vastly more among the Fixed Stars (to say nothing of the noblest or invisible Parts of the Creation, nor of the particular Phænomena here below) as is sufficient to make us cry out with the *Psalmist*, *O Lord how manifold are thy Works! In Wisdom hast thou made them all!*

PART III.

The Truth of the foregoing SYSTEM *briefly* Demonstrated.

IN order to let the Reader see the Certainty of our present System of Astronomy, and to prepare the way for his entire Satisfaction, as to the noble Inferences that shall hereafter be drawn from the same, I shall now attempt, not only to prove the foregoing System, in all its Parts, to be very *probable*, and so preferable to any other Hypothesis; but to *Demonstrate* it to be *really true and certain*; and this after so familiar a manner, that ordinary Mathematicians may easily apprehend the Force of each Argument, and see the Evidence for the several Conclusions all along. Now the Propositions I shall here Demonstrate are these:

I. That the *Diurnal Motion*, or that which occasions the Succession of Day and Night, and the apparent rising and setting of the Sun, Moon, and Stars, in the space of 24 Hours, which we

call

call a *Day*, belongs to the Earth, and not to the Heavens.

II. That the *Annual Motion*, or that which occasions the Succession of Summer and Winter, and the apparent Motion of the Sun through the Ecliptick in the Space of $365\frac{1}{4}$ Days, which we call a *Year*, belongs to the Earth, and not to the Sun.

III. That there is an Universal *Power of Gravity* acting in the whole System; whereby every Body, and part of a Body, Attracts and is Attracted by every other Body and part of a Body through the whole System; that this Power of Gravity is greater in greater Bodies, and lesser in lesser; and this in the exact proportion of such their Magnitude: That it is also greater when the Bodies are nearer, and lesser when they are farther off; and this in the exact duplicate proportion of such their nearness: That this Power is the same in all Places, and at all Times, and to all Bodies: And that, lastly, this Power is entirely Immechanical, or beyond the Power of all material Agents whatsoever.

IV. That the *Orbits, Revolutions, Distances, Quantities of Matter, Densities, Gravity on the Surfaces, Revolutions about their Axes, Quantities of Light and Heat*, &c. above set down, concerning the Sun and Planets, both Primary and Secondary, with those of the Comets also, are true and certain; with an Account of the Ways whereby we discover every one of those Particulars.

PROP.

PROP. I.

The *Diurnal Motion*, or that which occasions the Revolution of Day and Night, and the apparent rising and setting of the Sun, Moon, and Stars, in the Space of 24 Hours, which we call a *Day*, belongs to the Earth, and not to the Heavens.

DEMONSTRATION.

(1.) All the Phænomena or Appearances relating to this Matter, are now certainly known to be equally natural and necessary Consequences of a diurnal Revolution of the Earth from *West* to *East*, as of the like Revolution of the whole System of the Heavens from *East* to *West*; as those who have apply'd themselves to this part of Astronomy do well know: Just as it is equal to a Person that desires to see quite round a Terrestrial Globe, set in a proper Position; whether he walks himself round that Globe, while it stands still; or whether the several parts of that Globe be turned round to him, while he stands still: Which if it be granted; and the Vastness of the System about us be consider'd, with respect to the Smallness of our Earth; the immense swiftness necessary in one case, compar'd with the greater slowness in the other; the prodigious diversity of perplex'd Motions in the Sun, the Planets, the Comets, and the fix'd Stars to be provided for in the former Hypothesis, with the easy simple Motion of one Globe about its own Axis to be allow'd in the latter; the disproportion, as to the probability of the one

one and the other Notion, will appear vastly great and prodigious. Whether is it more fit and reasonable for 100 Auditors, in a Course of Geography, to have a Terrestrial Globe turned once round on its own Axis, in order to their distinct and gradual view of the particular Countries thereon describ'd; or to have Carpenters set to work to remove the Room, and the House it self, and to carry it on Wheels in order to avoid that single Circumvolution? When once we are satisfyed of the Justness of such a procedure in the one Case, we may begin to think of allowing the like Justness in the other; but not sooner.

(2.) There are no mechanical Laws of Motion known in the World, which can account for such a Diurnal Revolution of the Heavens; nay, it is directly contrary to all such known Laws whatsoever. 'Tis true, a Clock or Machine may have several distinct Motions within, and yet a Spectator may turn the whole round on its Axis at the same time; because all the Parts and Wheels are connected together, and take hold of each other, by material contract and insertion: So that he who removes one part, does of necessity remove all the rest. But this is far from the real Case in the World about us; where the several Bodies are vastly remote from, and unconnected with each other; and where therefore no such (imaginary) Revolution of any (fictitious) *Primum Mobile*, or material external Sphere can affect or move the several Bodies therein contain'd. When once we see the Revolution of a large Wheel make other inward Wheels, which it does not touch, dance attendance thereto, and commence circular Revolutions

tions round its Axis, we may begin to think of such a parallel Notion as to the present System, but not sooner.

(3.) There is no Example of such Periodical Revolutions, which carry different and seperate Bodies round the same Axis, in the same time, in the whole World about us. No Vortices or Whirlpools excited in Fluids do so: None of the Primary Planets are carried so about the Sun, nor Secondary ones about their Primaries. Nor indeed is the thing possible in any mechanical Method whatsoever, agreeing to the present System of things, that we know of. So that 'tis meerly an Hypothesis or Romance, unsupported by all good Evidence, and deriv'd entirely from the Prejudices and Notions of the Vulgar, before they are acquainted with the Principles necessary to make them competent Judges in such Matters.

(4.) The frequent, if not constant, Diurnal Revolutions of the rest of the Heavenly Bodies, renders it most highly rational, if not necessary, to allow the like Revolution to our Earth. If we cast our Eyes abroad, and use Telescopes to assist them, we shall find that *Jupiter* and *Mars* among the higher Planets; that *Venus* among the lower; that the Moon in our Neighbourhood; and that the Sun it self in the Center of these Planetary Motions, have, for certain, such a diurnal Revolution about their own Axes. Nor is it any way certain, that either *Saturn*, or *Mercury*, the secondary Planets, the Comets, or fixed Stars, *i. e.* that any of the Heavenly Bodies are destitute of such a Motion. So that hence it is exceeding probable that our Earth may have the like Motion also.

(5.) The

(5.) The contrary Hypothesis introduces the utmost Confusion into the apparent Heavenly Motions, while 'tis certain there is no manner of reason for doing so. For what strange Confusion is it to imagine, that while in *Venus*, in the *Moon*, in *Mars*, and in *Jupiter*, if not in all the other Planets and heavenly Bodies besides, there is such a diurnal Rotation, as severally produces a regular succession of Day and Night in them, according to the several natural Periods of those Revolutions; and while we cannot deny that the like diurnal Revolution of the Earth, would regularly produce a correspondent regular Succession of Day and Night in 24 Hours with us, without the introduction of any farther Motions of the Heavens for this purpose; what Confusion, I say, is it after all this, for us to introduce a strange, a violent, an unexampled, an unphilosophical Circumgyration of the whole vast Universe about our poor Earth, every Day, to the disorder and perplexity of those other diurnal Appearances, and of the whole System? and all this without any just occasion in the World? If we were but for a while translated to *Jupiter*, which we know to have the quickest diurnal Revolution of all the rest, and observ'd how regular the rising and setting of the Sun, Moon, and Stars, appear'd therein from that diurnal Motion alone, I dare say we should never after that so much as dream of any other than a diurnal Motion of our Earth, to account for the like rising and setting of the same Sun, Moon, and Stars with us here upon Earth.

(6.) In Fact, our Earth certainly has such a diurnal Revolution about its own Axis, as we are now speaking of: For though we do not stand conveniently enough to *see* the diurnal Revolution as to our own Earth, which we do as to the other Heavenly Bodies; yet are we capable of certainly knowing by one grand Effect of such a Motion, whether our Earth has that Motion or not? All Globes which have no diurnal Revolutions about their own Axes, must, by the Equality of the weight of Bodies in all their Regions, be perfect *Spheres*; and all the Parts of their Surface must, generally speaking, be at the same distance from the Center. But all Globes that have such a diurnal Rotation, (which will necessarily be swiftest at the Equator, and by consequence will cause the Parts to recede from the Axis of Motion, chiefly near the Equator) will be *Oblate Sphæroids*, or higher in the Equatoreal, and lower in the Polar Regions, as has been already observ'd. Now to this certain κριτήριον do we appeal for the determination of this matter. For since it appears from the lesser length of the Pendulum which vibrates Seconds near the Equator, than near the Pole; that the Surface of the Earth is about 31 Miles higher at the Equator, than at the Poles; and since the regress of the Earth's Nodes, which we call the precession of the Equinox, with the Equilibration of the Waters near the Equator, and near the Poles, do both fully confirm the same thing; all which are the necessary Effects of the Earth's diurnal Rotation, and are accountable on no other Principles whatsoever; I conclude, that our Earth has such a diurnal Rotation; or, which is the same thing, that the diurnal Revolution belongs to the Earth, and not to the Heavens.

See Sir Isaac Newton's Princip. 2d. Edit. p. 337, 338. and p. 437, 438.

N. B. I

N. B. I propose the Five first as Arguments exceeding probable; but this last as a really certain Demonstration.

PROP. II.

The *Annual Motion*, or that which occasions the Succession of Summer and Winter, and the apparent Motion of the Sun through the Ecliptick, in the space of $365\frac{1}{4}$ Days, which we call a *Year*, belongs to the Earth, and not to the Sun.

DEMONSTRATION.

(1.) All the Phænomena or Appearances relating to this Matter, are now certainly known to be equally natural and necessary Consequences of an Annual Revolution of the Earth, as of the Sun; as all Astronomers confess. And he who considers the prodigious Greatness of the Sun's Body, and the comparative Smallness of the Earth, will be under no Temptation to suppose that the vast Sun revolves round this little Earth; especially when he reflects, that all things will be the very same, if this little Earth be suppos'd to revolve about that vast Sun in the same time.

(2.) There are no known Laws of Motion according to which so great a Body as the Sun, can revolve about so small a Body as the Earth; nay, this is directly contrary to all such known Laws whatsoever. For let the Occasion or Influence derived from these two Bodies be of what sort you please, either Gravitation, or Magnetism, or Impulse, *&c.* still the greater Body,

Body, as in all like Cases, must have the greatest Force and Efficacy to move the other. And we may full as well expect that a Sling, containing a Milstone in it, may be fasten'd to a Pebble, and continue its motion about that Pebble, without removing it, as that the Sun can revolve about our Earth, while the Earth continues immoveable in the Center of that Motion.

(3). There is no Instance of such a thing in the visible World, as a great Body thus revolving about a small one, that continues it self immoveable all the while; but all the Instances are on the other side. The smaller Moon revolves about the larger Earth; the smaller Circumjovials and Circumsaturnals revolve about the larger *Jupiter* and *Saturn*; *Saturn*, *Jupiter*, *Mars*, *Venus*, and *Mercury*, revolve all, not about the smaller Earth, but about the larger Sun; as is now confess'd by the whole Astronomical World; and as is certainly demonstrated by the Phases of *Venus* and *Mercury* seen through a Telescope. And tho' in a strict sense all these larger Bodies may be so far said to revolve about those smaller, as they may still, on both sides, revolve about the common Center of Gravity; yet because the common Center of Gravity of the Sun and Earth must be so very near the Center of the Sun, and so very far from the Center of the Earth; and by Consequence, the Motion of the Sun, if compar'd with that of the Earth about it, must be so very insensible; it follows, that the most sensible Annual Motion, of which we are now speaking, must still, by all parallel Instances, belong to the Earth, and not to the Sun.

See Fig. I.

(4.) The

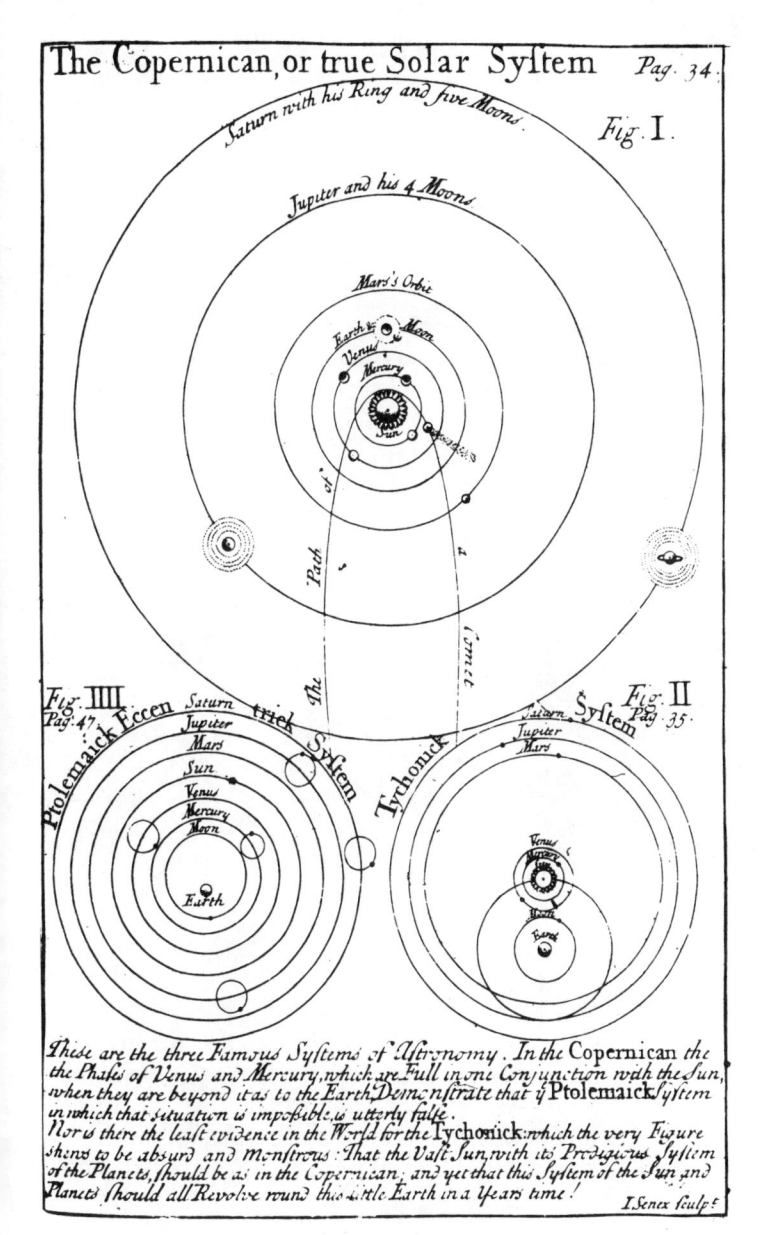

The Copernican, or true Solar System
Pag. 34.

Fig. I.

Saturn with his Ring and five Moons.

Jupiter and his 4 Moons.

Mars's Orbit

Earth
Moon
Venus
Mercury
Sun

the Path of a Comet

Fig. IIII. Pag. 47. **Ptolemaick Eccentrick System**

Saturn
Jupiter
Mars
Sun
Venus
Mercury
Moon

Earth

Fig. II. Pag. 35. **Tychonick System**

Saturn
Jupiter
Mars

Venus
Mercury
Sun
Moon

Earth

These are the three Famous Systems of Astronomy. In the Copernican the the Phases of Venus and Mercury, which are Full in one Conjunction with the Sun, when they are beyond it as to the Earth, Demonstrate that ye Ptolemaick System in which that situation is impossible, is utterly false.
Nor is there the least evidence in the World for the Tychonick: which the very Figure shews to be absurd and Monstrous: That the Vast Sun, with its Prodigious System of the Planets, should be as in the Copernican; and yet that this System of the Sun and Planets should all Revolve round this Little Earth in a Years time!

I. Senex sculp.t

of RELIGION.

(4.) The Hypothesis that this Annnal Motion belongs to the Sun, and not to the Earth, introduces the utmost Confusion among the Celestial Motions, and that without any proper occasion in the World. For what strange confusion is it for us to imagine that the single Earth, which is now confess'd by all to be situate between *Mars* above, and *Venus* below it; while all the rest of the Planets, *Saturn* with his *Satellites*, *Jupiter* with his, and *Mars*, *Venus*, and *Mercury*, without any, do all revolve not about the Earth, but about the Sun; nay that the Superior revolve about it in longer, and the Inferior in shorter Periods or Years, while the Annual Period, is in a mean between them; that I say the single Earth be exempted from the common Law of the whole System? What a heap of Absurdities are here? That while all the rest of the Planetary World revolves about the Sun, in their several regular Periods or Years, that yet our Earth, contrary to all Probability, should be suppos'd to carry not the Sun only, but all the Solar System also round it self in in a Year's time, as it certainly must, if the Annual Motion belong to the Sun? *See Fig.* II. This Hypothesis of the famous *Tycho*, which is the only one that is not absolutely impossible to be true, is yet so wild, groundless, and extravagant in it self, and so prodigiously improbable, that I should exceedingly wonder at its first Introduction, and much more at its Admission still in *Roman-Catholick* Countries, did I not know that Injudicious Persons have interpreted Scripture against the true System; and that an *Infallible Church* has Establish'd that Interpretation; nay, has condemn'd the true one as Heresy; and that, by Consequence, the Popish Astronomers have been driven into this terrible Dilemma, of either submitting to the Absurdities of the *Tychonian*, or else of resigning themselves up to Imprisonment, if not to Death it self for the true or *Copernican Hypothesis*: So dangerous is it, in that Church at least, if not in some others, to admit of Truth not only in Divine, but even in Natural and Astronomical Matters also!

(5.) There is one known fixed Proportion between the Periods and Distances of all the Heavenly Bodies from the Sun; I mean that the Squares of the Periods, are ever as the Cubes of the Distances; and this Proportion is so universal, that it obtains not only in this general System of the Primary Planets about the Sun, but in the particular Systems of the Secondary Planets about their Primaries, without the least Exception that we know of in the Universe: Which Proportion demonstrates, that the Annual Revolution belongs to the Earth, and not to the Sun. For as the Cube of the Moon's distance, which is 240,000 Miles, to the Cube of the Sun's distance, which is 81,000.000 Miles, *i.e.* as 1 to 38,272.753, so is the Square of the Moon's Period, or of less than 28 Days, which is about 784, to the Square of the Period of any Body moving about the Earth at the Sun's distance in Days, or to 30,005.838,352, whose Square Root, or 173,510 are therefore the Number of Days of the Solar Year, in case the Sun revolves about the Earth; which yet in reality are known to be but $365\frac{1}{4}$. Now this last Number is contain'd no less than 475 Times in the former. So that if the Sun revolv'd about the Earth, from

that

Fig. III.
Pag. 37.

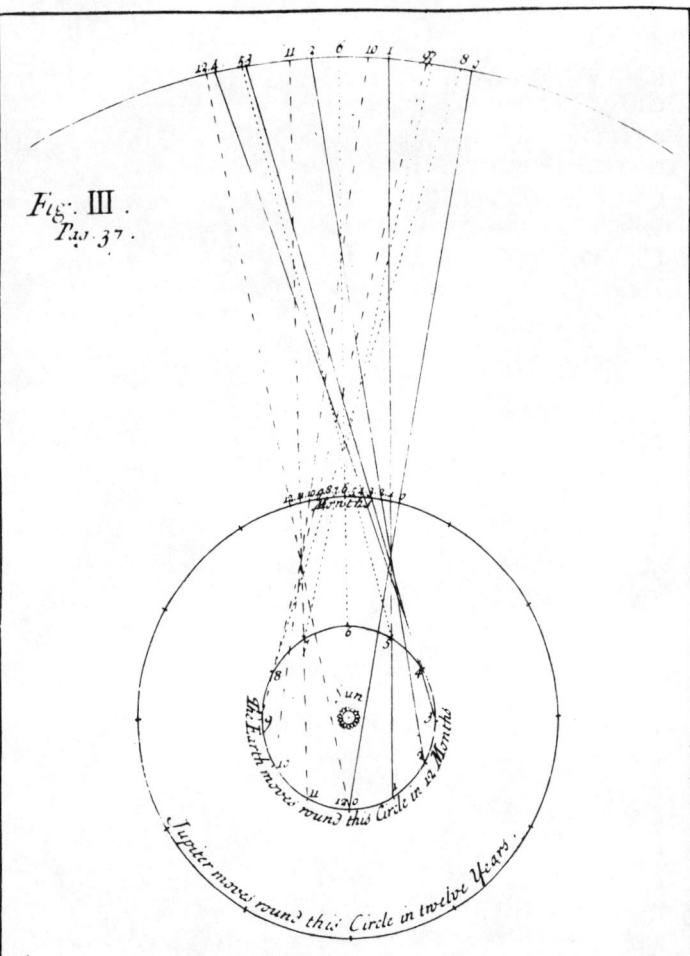

This Plate truly represents the Direction, Station, and Retrogradation of the Superior Planets, by the Situation of the Earth and Jupiter, one of those Planets, during a whole Years motion of the One, and a Months motion of the Other; from their first Position at 1.1.1 when Jupiter appears Direct, and near the Conjunction; at 2.2.2 and 3.3.3 when it continues Direct, tho' more slow in its motion; about 4.4.4 when it is Stationary; at 5.5.5 when it appears Retrograde, as also at 6.6.6 (when it is in Opposition) and 7.7.7; till about 8.8.8 it is Stationary again: After which at 9.9.9 and 10.10.10 it becomes Direct again, and more swiftly so at 11.11.11 and 12.12.12. The Case of Saturn and Mars are much the same.

I. Senex sculp.

that Analogy and Harmony which is every where else observed in the Heavens, the Year must be 475 Times as long as it now is. This Calculation seems to me to determine this Dispute in favour of the *Copernican* System; according to which, the Proportion above-mentioned of the Cubes of the Distances, as the Squares of the Periodick Times, is known to agree to all the Planets, to the utmost exactness of Astronomical Observations.

(6.) In Fact, the Earth has such an Annual Revolution about the Sun; as appears by all the Indications of such a Motion possible: even by that of Parallax; which is the known Geometrical Foundation of the Science of Astronomy. Thus if the Earth have an Annual Revolution about the Sun, it must affect the apparent Motions of all the other Planets and Comets, and, notwithstanding the Regularity of their several Motions in their own Orbits, must render the regular Motions, as to us, living upon the moving Earth, sometimes Direct, and that swiftly or slowly; sometimes Stationary, and sometimes Retrograde, and that swiftly or slowly also; and all this at such certain Periods, in such certain Places, for such certain Durations, and according to such certain Circumstances, as Geometry and Arithmetick will exactly determine, and not otherwise. Now that this is the real Case in Fact; and that every one of these Particulars are true in the Astronomical World, all that are skilful in that Science do freely confess; even those who, for Reasons already hinted at, do not think fit to declare openly for this Annual Revolution of the Earth, which is the natural, the certain Consequent of that Concession.

See Fig. III.

'Tis true, such Persons may pretend, that tho' these Phænomena be undoubted; and would be the undoubted Effects of such a Motion; yet that this Annual Revolution is not the undeniable, the strictly Geometrical Consequence of them; that they *may possibly* be accounted for on other Hypotheses, and that on the Supposition of certain complex Motions, deriv'd from (imaginary) Epicycles, and Eccentricks, and solid Orbs, &c. they may be solved without the introduction of this Annual Motion of the Earth. Now this I do not absolutely deny, that if such precarious, aukward, immechanical, false, and absurd Figments, were the true and real Laws of Nature, these Phænomena *might* possibly be otherwise accounted for: But then I affirm, that such Hypotheses are indeed not at all the real Laws of Nature, but no better than meer precarious, aukward, immechanical, false, and absurd Figments only; such indeed, as if they were allow'd in other Cases, would take away all Certainty in all mixt Mathematicks at least, if not in Arithmetick and Geometry themselves; and would enable Men to evade the grand Foundation of Astronomy, I mean that of the common Parallax it self: These Phænomena being as truly Instances of an Annual, as other known ones are of the Diurnal Parallax. I believe the skilful Astronomers will know my meaning by this general intimation; but if not, I will easily undertake to demonstrate, that those who, notwithstanding these Indications, or Demonstrations, do deny or doubt of the Earth's Annual Motion, may, in Consequence thereof, become Astronomical Scepticks, and deny, or doubt of almost all the other Principles of Astronomy,

and

and of Human Knowledge, which are never paſt the Evaſions of reſolved Scepticks. But I need not here enlarge, becauſe I am well aſſur'd that all Aſtronomers, who are compleat Maſters of that Science, as ſoon as other Prejudices and Fears are over, will agree with me, that the Arguments under this Head, when allow'd their free and full Weight, do certainly prove that this Annual Revolution belongs to the Earth, and not to the Sun. I do not here mention that Annual Parallax of the *Fixed Stars*, which Dr. *Hook* and Mr. *Flamſteed* think they have diſcovered, which would certainly *demonſtrate* the Earth's Annual Motion, and which I have elſewhere vindicated from the Objections of Dr. *Gregory*, and of the *French*; becauſe it is not yet generally allow'd for true by the Learned. Nor need I have recourſe to that Attempt: ſince the moſt evident Annual Parallax of the Planets and Comets, already inſiſted on, does plainly prove this Annual Motion of the Earth. *See Aſtron. Lect. IV. And Math. Philoſophy. Lect. XXI.*

N. B. The Reader will eaſily perceive, that I propoſe the former Five Arguments as highly probable, but this laſt as a certain Demonſtration of the Annual Motion of the Earth.

N. B. I do not here meddle with the ſeveral Objections made formerly againſt either the Diurnal or Annual Revolutions of the Earth, either from Scripture or from Nature; ſince there are few of the truly Learned and Judicious which do now inſiſt upon them; and ſince they have been fully conſider'd and confuted by others, in treating on this Argument; to whom I ſhall refer *See Galileo's Syſt. Coſmic. Mr. Derham's Aſtrol. Theol. Pref.*

refer the Reader, if he still want Satisfaction therein.

PROP. III.

(1.) There is an universal *Power of Gravity* acting in the whole System; whereby every Body, and part of a Body, Attracts and is Attracted by every other Body and part of a Body, through the whole System. (2.) This Power of Gravity is greater in greater Bodies, and lesser in lesser; and that in the proportion of such their Magnitude. (3.) It is also greater when the Bodies are nearer, and lesser when they are farther off, and that in the exact duplicate proportion of such their nearness. (4.) This Power is the same in all Places, and at all Times, and with regard to all Bodies whatsoever: (5.) This Power is entirely immechanical, and beyond the Abilities of all material Agents whatsoever.

Demonstration *of the* first Part; *That there is such a Power of* Gravity *in the Universe.*

(1.) Because all the Planets, Primary and Secondary, with the Comets, are perpetually drawn from their natural Rectilinear Courses along strait Lines, the Tangents of their present Orbits, and made to revolve in Curves; there is therefore a continual Power or Force acting upon them; and because the Power or Force acts so upon them, as to cause the Bodies to move just so much quicker, as they are nearer their Central Bodies; and just so much slower, as they are farther off them; and thereby to oblige them by a Line drawn from the Central

tral Bodies ever to describe equal Area's in equal Times; it therefore follows, by Geometrick Demonstration, that this Power or Force always tends directly to the Centers of those Bodies about which they revolve; or is still a Centripetal Power or Force, with regard to the same Bodies.

See Math. Philos. Prop. 16.

(2.) This Power or Force is properly the Power or Force of *Gravity*, or the very same which causes Stones and all heavy Bodies with us to *gravitate*, and thereby to descend downward towards the Earth's Center; because upon exact Geometrick Deduction, and Arithmetick Calculation it appears, that the Force which retains the Moon in its Orbit, tends to the same Center, and is exactly of the same Quantity with that, which accelerates heavy Bodies with us on the Earth's Surface; due Allowance being made for the Difference of their Distances from the Earth: which Force causes all Bodies whatsoever on the Earth's Surface to descend a little more than 16 *English* Feet, in one Second of Time.

See Math. Philos. Prop. 79.

Demonstration *of the* Second Part, *that this Gravity is exactly proportional to the Magnitude of the Central Body.*

(1.) Because the Heavenly Bodies are all Sphærical, and the Force of Gravity tends still directly to their several Centers, that Force, by Geometrical Demonstration, is the very same as if it tended to every Particle of those Central Bodies, and as if the Gravity to each whole Central Body were compos'd of the Gravity to each

See Math. Philos. Prop. 45.

each Part of the same; which indeed is the only rational Conception of this Matter; because otherwise this Force must respect not a Body, which is somewhat real, but a Mathematical Point, which, physically speaking, is nothing at all.

(2.) In Fact, the greatest Bodies have the greatest Gravity towards them through the whole Universe. Thus by Calculation from the Heavenly Motions it appears, that the vastly greatest Force of Gravity is that tending to the *Sun*, the next that to *Jupiter*, the next that to *Saturn*, the next that to the *Earth*, and the next that to the *Moon*; according to the known Order of the Magnitude of their several Bodies respectively. Nor do I omit the other Planets and the Comets here, as if the Case were not the same in them also; but because they have no Planets, that we know of, moving about them; and because they occasion no sensible Tides in our Ocean; which are the only Phænomena, whence we can draw any certain Indications of the Quantity of that Power of Gravity which tends to them.

See Math. Philos. Prop. 84.

Demonstration *of the* Third Part; *that this Gravity is exactly in the Duplicate Proportion of the Nearness of Bodies,* i. e. *when twice as near, four Times; when thrice as near, nine Times; when four times as near, sixteen Times as strong; and so for ever.*

(1.) Because the Cubes of the Distances of the several primary Planets about the *Sun*, and
of

of the several Secondary Ones about their Primaries, are still found to bear accurately the same Proportion to each other, which the Squares of their respective Periodical Times of Revolving also bear; it follows, by strict Geometrick Reasoning, that the Centripetal Force, which keeps them in their several Curves, is accurately in the duplicate Proportion of their Nearness to their Central Bodies. *See Math. Philos. Prop. 13.*

(2.) Because all the primary Planets and Comets revolve about the *Sun*, and all the secondary Ones, that are Eccentrical, about their Primaries in Ellipses, or Conick Sections, and that about the *Sun*, or their Primaries, situate in their *Foci*; it follows, by strict Geometrick Reasoning, that the Centripetal Force, which retains them in such Conick Sections, is exactly in the duplicate Proportion of their nearness to their Central Bodies. *See Math. Philos. Prop. 21.*

(3.) Because the *Aphelia*, or longer Axes of these Ellipses do rest, as to the fixed Stars, in the primary Planets and Comets; and in the secondary Ones, do move only in proportion to the disturbance they necessarily meet with from the Inequality of their Distances from the *Sun*, it follows from Mathematick Reasoning, that the Centripetal Force, which acts upon those Planets and Comets, is most accurately in the duplicate Proportion of their Nearness to their Central Bodies. *See Math. Philos. Prop. 33.*

N. B. That there appears no manner of Necessity for this Proportion of Increase and Diminution of the Power of Gravity, as the Squares of the Nearness of Bodies more, than for any other Proportion; as suppose that of
the

the Nearness it self, or that of the Cubes, or Biquadrates, &c. of that Nearness or Distance. Yet would any other Proportion have been exceeding inconvenient for this System. For if it had been as the Nearness it self, we have no regular Curve Line, in which the Eccentrical Planets or Comets could then revolve: If in a Triplicate such Bodies must ever have descended, or ascended in Spiral Lines, till they fell into the Sun, or some Fixed Star, and perished therein. And if the Proportion had been still greater, the Effect would have been quicker also.

See Math. Philos. Prop. 20.

Demonstration *of the* Fourth Part; *that this Gravity is the same in all Places, and at all Times, and with regard to all Bodies whatsoever.*

This is plain from Fact and Observation; for (1.) Heavy Bodies fall as fast in one Country as in another; and Gravity affects the Planets and Comets in its proper Proportion, equally in all the Parts of the Universe where Planets or Comets go, *i. e.* in all the immense Regions of it every Way.

(2.) The Planets and Comets, guided by this Law, have invariably preserv'd their Motions, Orbits, Periods, and Influences, since the first Ages of Astronomy and Chronology: And,

(3.) This Gravity does still, upon Tryal, equally affect Fluids and Solids; Bodies in Motion and at Rest; Great and Small; and this through the whole Universe also, so far as we can examine it.

Demon-

Demonstration of the Fifth Part; *that this Gravity is an entirely Immechanical Power, and beyond the Abilities of all material Agents whatsoever.*

(1.) This Power acts upon the very inward substantial Parts of Bodies, as well as the outward and visible; and is proportionable not to the Surface but to the solid Content, or Quantity of Matter contained in them: Whereas, all Mechanical Causes are meerly superficial, and act by External Contact on the External Surface only. *See Math. Philos. Prop. 82.*

(2.) This Power acts upon Bodies equally, when they are in the most violent Motion, and when they are at Rest; as the Celerity of Descending Bodies with us; and the Celerity of the Comets in the Heavens, Geometrically computed, do particularly shew. Now this is absolutely impossible; that any Mechanical Pressure, or Impulse from a Body, let its Motion be never so swift, or its Pressure never so strong, should equally accelerate another Body, when at Rest, and when in Motion; it being a known Law of Mechanism, that a Body in Motion impels another at Rest with its whole Force; but one in Motion, with only the Excess of its own Velocity above the others; as is most obvious also on the least Reflexion. *See Math. Philos. Prop. 3.*

See Math. Philos. Law of Motion 6, &c.

(3.) By this Power Bodies act upon other Bodies *at a Distance*, nay at all Distances whatsoever; that is, they *act where they are not*: Which is not only impossible for Bodies Mechanically

nically to do, but indeed is impossible for all Beings whatsoever to do, either Mechanically or Immechanically, it being just as good Sense to say, an Agent can act *when* he is not in Being, as *where* he is not present. Whence, by the way, as we shall see hereafter, it will appear, that this Power of Gravity is not only Immechanical, or does not arise from Corporeal Contact or Impulse, but is not, strictly speaking, any Power belonging to Body or Matter at all; tho' for ease of Conception and Calculation we usually so speak; but is a Power of a superior Agent, ever moving all Bodies after such a manner, as if every Body did Attract, and were Attracted by every other Body in the Universe, and no otherwise.

N. B. Altho' I here do only insist in particular on the wonderful Power of Gravity, which is the general or universal Power of the entire System; and which is the best known, the most easily proved, and is indeed the most evidently Immechanical; yet do not I exclude those other Noble and Immechanical Powers of Refraction in pellucid Bodies; of mutual Repulse in the Particles of the Air, which render it Elastical; of the Cohesion of Parts in consistent Bodies, and of another Kind of Attraction in homogenial, and of Repulse in heterogeneous Fluids, *&c.* on which the particular Phænomena of Nature do now appear to depend; but shall upon Occasion make use of them sometimes in what follows.

P R O P.

PROP. IV.

The Orbits, Revolutions, Distances, Diameters, Quantities of Matter, Densities, Gravity on the Surfaces, Revolutions about their Axes, Quantities of Light and Heat, above set down, concerning the Sun and Planets, both primary and secondary, with those of the Comets also, are True and Certain. With an Account of the Ways whereby we discover all those Particulars.

Demonstration of the several Particulars.

(1.) That the *Orbits* of all the primary Planets are, as above stated, *Elliptical*, was first discovered, and prov'd with the utmost Labour and Industry, from *Tycho*'s Observations, by the famous *Kepler*; particularly as to the Planet *Mars*; and is now discover'd by the great Exactness of all the Planets Places, and apparent Diameters, when calculated in such Orbits, and compared with the best Observations. Nor was the old Hypothesis of the *Eccentrick Circular Orbits*, other than an Approximation to the true *System*; especially when the Astronomers came to observe, that their Circular Eccentricity would not agree to Observation, without its *Bisection*, as it was call'd; *i. e.* without supposing the Point of even equable Motion to be as far *beyond*, as the Eccentrical Point was on *this side* the Center of the Circle, which was almost the same Thing with owning such Orbits to be Elliptical, and having the Sun in the Inferior,

De Motibus Stellæ Martis.

See Fig. IV.

rior, and the Point of even Motion in or very near the superior Focus; which is the State of those Orbits in our present Astronomy. And the Case will be the same as to the Orbits of the Secondary Ones about their Primaries, where-ever their Eccentricity can be discover'd. Nor is the State of the Comets to be excepted from this Rule: For tho' We now generally use Parabolick instead of Elliptick Orbits in the Computation of the Comets Motions, yet is this only for that small part of their Course which we can see, and for the ease of Calculation; while we are satisfy'd from their Appearances, that they really move in very Oblong Ellipses, (or such as approach to Parabola's,) about the Sun in their lower *Foci*, as do the Planets; that accordingly they have their proper Periods, in which they revolve quite round, and return to the Sun again; three of which Periods are now discovered; that if we compute them in Ellipses, instead of Parabola's, where-ever the Species of those Ellipses are known, we should find our Calculations still more exactly to agree with our Observations, as is already the Case as to the Comet 168^0_1.

See Math. Philos. of Comets, P. 381, *&c.*

(2.) The *Revolutions* of the primary Planets about the Sun, and of the Secondary Ones about their Primaries are above by me thus stated.

		D.	H.	.
Mercury	revolves	87	23	16
Venus	about the	224	16	49
The Earth	Sun in	365	6	9
Mars	the Space	686	23	27
Jupiter	of	4332	12	20
Saturn		10759	6	36

Circumjovials.

		D.	H.	.
The Innermost	revolves a-	1	18	28
The Second	bout *Jupi-*	3	13	14
The Third	*ter* in the	7	3	43
The Fourth	Space of	16	16	32

Circumsaturnals.

		D.	H.	.
The Innermost	revolves	1	21	19
The Second	about *Sa-*	2	17	41
The Third	*turn* in	4	13	47
The Fourth	the Space	15	22	41
The Fifth	of	79	22	4

	D.	H.	.
The Moon revolves round the Earth, from any fixed Star, to the same again, in	27	7	43
From the Sun to the Sun again, in	29	12	44

	Y.	M.	D.
First of the Three Comets revolves in	75	0	0
Second,	129	0	0
Third,	575	0	0

This is plain and direct Matter of Observation; and depends only on the Comparison of the former Places of Planets noted by Astronomers, to the fixed Stars, to the Sun, or to their Primaries, with the latter; and the easy Computation of the Number of Years, Days, Hours, and Minutes, between such Observations. Thus, for Instance, we first know, by gross Observation, that the Moon is above 27 Days in revolving round us, from any fixed Star to the same again; we then compare several Months at once, and find that in 12 Months one with another, the Moon is almost Eight Hours above 27 Days in such a Revolution. And then, lastly, we compare the like Revolutions for many Years together, and so discover that just 27 Days, 7 Hours, and 43 Minutes, is its true Periodical Time.

See Fig.   *N. B.* The ordinary or Diurnal Parallax of any Celestial Body, is the small Angle that is made in a plain Triangle composed of Three strait Lines; the One from the Center of the Body, to a Spectator on the Earth's Surface: The Second from that Center to the Earth's Center: The Third from the Spectator to the Earth's Center; or that Angle which is ever over against the Earth's Semidiameter. And further,

N. B. If the Reader be unacquainted with plain Trigonometry, he is to observe, that in every plain Triangle, if Two Angles be known, the Third is of Course known; and the *Proportion* of the Sides is known also: This he may try himself in any Triangles equiangular to one another, and may, by measuring the Sides, see that their *Proportion* is perpetually the same.

(3.) The

Fig. V
Pag. 50

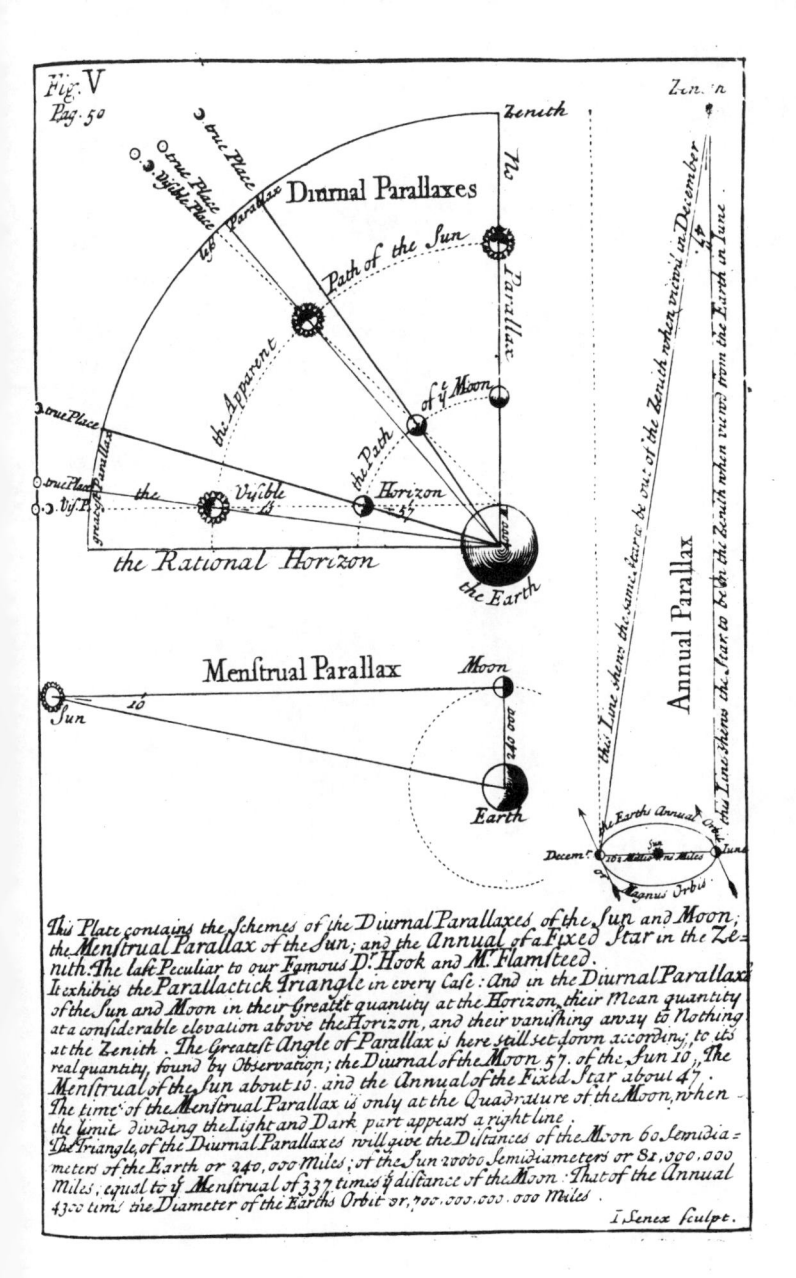

Diurnal Parallaxes

Zenith

☽ true Place
☽ Visible Place
true Parallax
Path of the Sun
No Parallax
of ☽ Moon
the Apparent
true Place
true Place Vis. Pl.
the Visible ☉ the Path Horizon
the Rational Horizon
the Earth

Menstrual Parallax

Sun — 10 — Moon
240000
Earth

Annual Parallax

Zenith

this Line shews this semidiameter to be out of the Zenith when viewed in December
this Line shews the Star to be in the Zenith when viewed from the Earth in June

the Earth's Annual Orbit
Decemr. 190 Mil. ☉ Sun 190 Mil. June
Magnus Orbis

This Plate contains the Schemes of the Diurnal Parallaxes of the Sun and Moon; the Menstrual Parallax of the Sun; and the Annual of a Fixed Star in the Zenith. The last Peculiar to our Famous Dr. Hook and Mr. Flamsteed.
It exhibits the Parallactick Triangle in every Case: And in the Diurnal Parallax of the Sun and Moon in their Greatest quantity at the Horizon, their Mean quantity at a considerable elevation above the Horizon, and their vanishing away to Nothing at the Zenith. The Greatest Angle of Parallax is here still set down according to its real quantity, found by Observation; the Diurnal of the Moon 57′. of the Sun 10″. The Menstrual of the Sun about 10′. and the Annual of the Fixed Star about 45″. The time of the Menstrual Parallax is only at the Quadrature of the Moon, when the limit dividing the Light and Dark part appears a right line.
The Triangle of the Diurnal Parallaxes will give the Distances of the Moon 60 Semidiameters of the Earth or 240,000 Miles; of the Sun 20000 Semidiameters or 81,500,000 Miles, equal to ye Menstrual of 337 times ye distance of the Moon. That of the Annual 4300 times the Diameter of the Earths Orbit or 1,700,000,000,000 Miles.

I. Senex Sculp.

of RELIGION.

(3.) The real mean *Distances* of the Primary Planets, and the known Comets, from the Sun, as also of the Secondaries from their Primaries, have been by me thus determined.

Mercury		32,000.000
Venus		59,000.000
The *Earth*	is distant from the *Sun's* Center about	81,000.000
Mars		123,000.000
Jupiter		424,000.000
Saturn		777,000.000
Nearest of the 3 Comets		1.458,000.000
Middlemost		2.025,000.000
Outmost		6.600,000.000

Statute Miles; each 5280 *English*, and 4943 *French Feet*.

Circumjovials.

The Innermost		2,300.000	
The Second	is distant from the Center of *Jupiter* ab^t	368.000	Miles.
The Third		580.000	
The Fourth		1,000.000	

Circumsaturnals.

The Innermost		146.000	
The Second	is distant from the Center of *Saturn* ab^t.	187.000	
The Third		263.000	Miles.
The Fourth		600.000	
The Fifth		1,800.000	

The Moon { is distant from the Earth's Center about } 240,000 { Miles.

These Distances, as to their Proportion, are well known from plain Trigonometry; I mean, by the utmost Elongation of the Inferiors; where in a Rectangular Triangle, joining the Eye, the Center of the Sun, and the Center of the Planet, two of the Sides bear this Proportion to each other: And by the Angle of Retrogradation, thereto equivalent, in the Superiors; for 'tis but imagining your Eye, transferr'd to the Superior Planet, and the Earth is an Inferior one with respect to the Eye, and the Proportion as before is given; as also by the Position of *Jupiter*'s Shadow, in the Eclipses of its *Satellits*, where the Middle of the Eclipse gives the Position of the Central Shadow of *Jupiter* from the Sun; and our Instruments give the Position hence: So that we have all the Angles in a plain Triangle made by the Center of the Sun, the Center of *Jupiter*'s Shadow, and the Eye, which gives the Proportion of the Sides or of the Distances; and by the Proportion of the Periodick Times, compar'd with the mean Distances in the Comets, all which Methods agree together in the present Case. And as to the Distances in Miles themselves, they are derived from the best Observations of the Parallax of *Mars* and *Venus*, by Mr. *Flamsteed*, *Cassini* and others, which give us that of the Sun about 10″, and which is followed by Sir *Isaac Newton* in the second Edition of his *Principia*. Only we may Note, that this Parallax cannot be much

much larger, because that would easily be discover'd by our Instruments, but may be a little less, because that is more difficult to be found out. So that it is more probable that these Distances are somewhat too little, than that they are at all too great. But as to such farther Exactness, it must be left to the nicest Observations of Posterity. Nor will it be at all difficult in that Case to settle more exact Numbers for these Distances before us; *viz.* those in the reciprocal Proportion to that Parallax.

Corollary. From these two Tables compar'd together, we learn the Mean Annual Horary Velocities of the Primary Planets, and of the Three Comets about the Sun, and of the Secondaries about their Primaries; as follows:

	moves in the Space of one Hour about	Miles
Mercury		100.000
Venus		70.000
The Earth		56.000
Mars		45.000
Jupiter		24.000
Saturn		18.000
Nearest of the 3 Comets		13.000
Middlemost		9.000
Outmost		6.000

N. B. Light moves about 650,000.000 Miles in an Hour; 10,800.000 in a Minute; and about 180,000 in a second of Time.

Circumjovials.

The Innermost	moves in the	32.700	
The Second	Space of	25.700	Miles.
The Third	one Hour	20.000	
The Fourth	about	15.000	

Circumsaturnals.

The Innermost		19.400	
The Second	moves in the	18.500	
The Third	Space of one	14.300	Miles.
The Fourth	Hour about	9.400	
The Fifth		5.600	

The Moon moves in the Space of one Hour about 2.200 Miles.

(4.) The real *Diameters* of the Heavenly Bodies have been by me thus determined, in *English* Statute Miles, according to Mr. *Flamsteed*'s Observations.

The Sun's	763.000
Saturn's	61.000
Jupiter's	81.000
Mars's	4.440
The Earth's	7.970
The Moon's	2.170
Venus's	7.900
Mercury's	4.240

These

These Numbers, which are discover'd by the apparent Diameters, compared with their real Distances, are as to the Earth and Moon, within a small Latitude, certainly true in the *Miles themselves*; and as to the rest, are within no great Latitude certainly true, in the *Proportion* of one to another; but not so certain as to the Miles themselves. The Reason is plain, that the real Bigness of the Earth, and of our Neighbour the Moon, are easily discover'd, and measur'd by the Rules of Astronomers; but that the rest are so vastly remote from us, that our nicest Instruments cannot yet perfectly define the *Distances* of any of them; and so by Consequence cannot perfectly determine their real *Diameters*, which depend thereon. Only so far I may, as before, venture to conjecture, that the real Distances and Diameters are rather larger than smaller than those here set down; because if the Distances, and so the Diameters, were much less, they would be certainly discover'd by our Instruments; whereas if they be suppos'd considerably greater, it must be still harder to discover those Distances and Diameters by the same Instruments.

N. B. The Diameters of the Circumjovials and Circumsaturnals, have not yet been exactly observ'd by Astronomers; so they can have no Place in this and some other Calculations. Only so far we perceive by Mr. *Huygen*'s Information, that they are full as large as our Moon; nay rather, as the lesser of the primary Planets themselves. *Cosmoth. p. 101.*

N. B. Hence we learn the Contents of all these Surfaces in Square Miles, by the help of the Elements of Geometry.

The *Sun* contains about	In Surface.	
	1,813200,000,000	
Saturn - -	14,000,000,000	
Jupiter - -	20,000,000,000	
Mars - - -	60,000,000	Square Miles.
The Earth - -	200,000,000	
The Moon - -	14,000,000	
Venus - -	200,000,000	
Mercury - -	55,000,000	

N. B. Hence also, by the same Help of the Elements, we learn the solid Contents of all these Bodies in Cubical Miles, as follows.

	In Solidity.	
The *Sun* contains about	230,000,000,000,000,000	
Saturn - - - -	160,000,000,000,000	
Jupiter - - - -	280,000,000,000,000	
Mars - - - - -	44,000,000,000	Cubical Miles.
The Earth - - - -	266,000,000,000	
The Moon - - - -	5,000,000,000	
Venus - - - -	264,000,000,000	
Mercury - - - -	39,000,000,000	

N. B. Mr. *Huygens*'s Numbers of the apparent Diameters, are considerably different from those of Mr. *Flamsteed*. And I do suppose still, what I once propos'd to the Reverend Mr. *Derham*, and which he approves of in his Astro-theology, that the middle Number between Mr. *Flamsteed*'s and Mr. *Huygens* may be the truest: Yet do I, in all my Calculations here, follow those of Mr. *Flamsteed*; because Sir *Isaac Newton* has done so, even in the last Edition

Edition of his *Principia*. When once new Trials have determin'd which of these Observations are most exact, it will not be very difficult to correct those Errors which arise from the present Uncertainty there is among the Observations to the present Diameters.

N. B. If we sum up the last Quantity of Matter for the Sun and Planets in the Solar System together; and add to it, for the Comets, another Quantity equal to that of all the Planets; and, allowing nothing for the interspers'd Vacuities, which yet are much the greatest Part, compare the whole with the Cubical Content of this System upon the Foot of Mr. *Flamsteed*'s Parallax of the fixed Stars, we shall find that the Quantity of the Matter is about 232,000,000,000,000,000 Cubical Miles; and that a Cube of the Diameter of the System, which I esteem equivalent thereto, is 1,000,000,000,000,000,000,000,000,000,000,000 of the same Miles, *i. e.* it will appear that this Quantity of Matter is little more than the 4,000,000,000,000,000,000.oooth Part of the empty Space; or in other Words, that a small Pin-head bears a much greater Proportion to a Cubical Mile, than all the Matter in this System bears to the empty Space therein contain'd.

(5. The *Quantities of Matter* in such of the Planets as afford us the Opportunity of discovering the same, have been by me thus stated;

    The *Sun*'s ---------- 227500
    Jupiter's ------------- 220
    Saturn's -------------- 94
    The *Earth*'s ------------ 1
    The *Moon*'s between --- $\frac{1}{39}$ and $\frac{1}{40}$.

These

These Numbers are agreeable to Sir *Isaac Newton*'s laſt Calculations, and belong only to ſuch of the Heavenly Bodies as have Planets about them, or which influence our Tides; from the Quantity of which Effects alone we can collect the Quantity of the Cauſes, or of the Matter in each Body. The way of diſcovering this Quantity is eaſy, from what has been already proved, that the univerſal Power of Gravitation is in proportion always to the Quantity of Matter in each attracting Body. For if from the conſtant Equality of Proportion there is between the Cubes of the Diſtances and the Squares of the Periods, we, by the Rule of Three, reduce the ſeveral Syſtems of the Planets to any one Diſtance from their Central Bodies, and there Geometrically compute the Proportion of the Centripetal Power to each Central Body, which in that Caſe is directly proportional to the Square of the Velocity of the Planet, we ſhall have the Quantity of the Force of Gravity towards every one of them; and, by Conſequence, the Quantity of that Matter which occaſions the ſame, and is proportionable thereto. For Example; Let us reduce the Diſtance of *Jupiter*'s innermoſt Planet from *Jupiter*, to the ſame Diſtance as the Earth has from the Sun; and obſerve the Difference of their Periods or Velocities in thoſe equal Circles, by this Rule; As the Cube of the *Satellits* real Diſtance 230,000 = 12,167,000,000,000,000, to the Cube of its imaginary Diſtance 81,000,000 = 531,441,000,000,000,000,000,000; ſo is the Square of its real Period in Minutes 2548, at its preſent Diſtance from *Jupiter* 6492304, to 283576603112024 the Square of its imaginary Period at the

ſame

of RELIGION. 59

same Distance from it, as the Earth has from the Sun; whose Square Root therefore 16,829.733 is the Period of that *Satellit* there in Minutes. Now as the Squares of these Periods reciprocally, which is as the Squares of their Velocities directly, *i. e.* as 276643388961 to 283586607511289, or as 229 to 227500 nearly, so are their Attractions towards their respective Central Bodies, and so are the Quantities of Matter in those Central Bodies; which was to be demonstrated: and so for the rest. By this Method we know the *Sun*'s, *Jupiter*'s, *Saturn*'s, and the *Earth*'s, Quantity of Matter. As for the *Moon*, this Method cannot reach it; because there are no Planets revolving about it. But then the Tides in our Ocean, arising from the *Sun* and *Moon*'s Influences, or the Water's Gravitation towards them; and the Spring Tides being the Effect of the *Sum* of their Forces, when they are united; while the Nepe Tides are owing to the *Difference* between them when they are opposite; we gather from the Observations made of those Spring and Nepe Tides, what the Proportion of their Powers is at their different Distances; whence we compute what it would be at equal Distances; *i. e.* what Proportion there is between the Quantity of their Matter respectively. Thus for Example; the highest Elevation of the Tide by the *Sun* and *Moon*, at the Conjunction and Opposition, is to its Elevation by the prevalence of the Moon over the Sun after the Quadratures, by the best Observations, as $5\frac{1}{2}$ to $3\frac{1}{2}$. Whence it follows, that the Moon's Force is to the Sun's as $4\frac{1}{2}$ to 1. For half the Sum of any two Quanties, added to half their difference giving the larger, and substracted from it gi-

See Sir Isaac Newt. Princip. 2d Edit.

ving the smaller, as any one may easily try, it follows from the proportion of the *Sum*, to the *Difference* above stated, $5\frac{1}{2}$ to $3\frac{1}{2}$; that their separate Forces must be in the proportion just now mentioned of $4\frac{1}{2}$ to 1. Whence that Quantity of Matter in the Moon, which at its proper distance will bear this proportion of $4\frac{1}{2}$ to 1 in the Elevation of our Tides and no other, must be its true Quantity. By a Computation from which Principles, it appears that the Moon has little more than the 9,100.000th part of the Quantity of Matter in the Sun; or, which is all one, little more than the 40th part of the Quantity in the Earth, as this Table informs us. For if you suppose the Sun as near as the Moon, its Force on the Tides will be so much greater than that of the Moon, as its Magnitude is greater; *i.e.* as about 9,100.000 is to 1, but then

See Math. Philos. *Prop.* 38. *and pag. prius.*

that Force being diminished as the Cube of the Sun's greater Distance is increas'd, or nearly as the Cube of $337 = 38,272.753$ to 1; is upon the whole so much more diminish'd than increas'd, that it amounts to only the $4\frac{1}{2}$th part of it; that being nearly the quotient of 38,272,753 divided by 9,100.000. By this means we are able to add the Moon's Quantity of Matter to that of the rest of the foregoing Bodies. But since neither *Mars*, *Venus*, *Mercury*, nor any of the Comets, have any visible Planets about them; nor do they sensibly affect our Tides; we have no means of knowing the Quantity of Gravitation towards them, and so no means of knowing the Quantity of Matter contained in them.

(6.) The *Densities* of the foregoing Heavenly Bodies have been by me thus already stated.

of RELIGION. 61

  The Moon's as $123\frac{1}{2}$
  The Earth's as   100
  The Sun's as $025\frac{1}{2}$
  Jupiter's as   019
  Saturn's as   15

These Numbers are thus discover'd. By what has been already set down, we know the real quantity of Matter in these several Heavenly Bodies; as also their true Diameters, and thence their entire Bulk or Magnitude; whence we having given quantities of Matter in given Spaces, we must have withal the Densities of the same quantities also.

Or thus; Spheres of the same Diameters with these Bodies, will not have their just quantities of Matter above stated, unless their Densities be as those Numbers here set down; whence it follows, that these and no other express their real Densities respectively.

Thus for Example: The Solid Space in the Earth is to that in the Moon, as the Cubes of their Diameters, or as $48\frac{1}{2}$ to 1; but, as we have seen, the quantity of Matter in the Earth is to that in the Moon only, as $39\frac{1}{2}$ to 1; whence it follows, that the Density of the Moon compensates its smallness, and is to that of the Earth as $48\frac{1}{2}$ to $39\frac{1}{2}$, or as $123\frac{1}{2}$ to 100, according to the Table before us; and the Case is the very same in the rest.

(7.) The *Weights of equal Bodies* on the *Surfaces* of the Sun and Planets last mentioned, have been by me thus determined.

On

On the Surface of the *Sun*, as   24,40
                 The *Earth*, as   1,00
                 Jupiter, as   2,00
                 The *Moon*, as   0,34
                 Saturn, as   1,28

That is, one Pound with us weighs on the Sun's Surface $24\frac{4}{10}$ Pounds, and so of the rest.

These Numbers are easily deriv'd from some of those that go before. For since we have already obtain'd those that express the Gravitation towards these Heavenly Bodies, at equal distances from their respective Centers; and since we have also already obtained their true Diameters, and thence we know their true Semi-diameters; and since we know withal, that the Power at different distances is ever as the Squares of those distances reciprocally: It cannot be difficult thence to compute the quantity of this Power at the particular distances of every ones Semi-diameter; which is the same with the Weight of equal Bodies on their respective Surfaces.

Thus for Example: The quantity of Matter in the Earth is to that in the Moon as $39\frac{1}{2}$ to 1, and in the same proportion do all Bodies gravitate to them at equal distances from their respective Centers. But since the Force of Gravitation diminishes, as the Square of the distance increases, and the distance of the Earth's Surface from its Center is to that of the Moon's from its Center, as 365 to 100, whose Squares are as $13\frac{1}{3}$ to 1, this greater distance diminishes the former Excess of Proportion, and reduces it to that of $39\frac{1}{2}$ to $13\frac{1}{3}$, or, as in the Table, to that

of RELIGION. 63

of 100 to 34. And the Case is the same in the rest.

(8.) The *Diurnal Revolutions* of the Sun and Planets about their own Axis, with respect to the Fixed Stars, have been already stated thus:

	D.	H.	.
The *Sun* revolves in about	25 :	6 :	0
Jupiter in	00 :	9 :	56
Mars in	1 :	0 :	40
The *Earth* in	0 :	23 :	56
Venus in	0 :	23 :	00
The *Moon* in	27 :	7 :	43

The way by which these Diurnal Revolutions are discovered in the *Sun, Jupiter, Mars*, and *Venus*, is obvious; I mean the Observations by Telescopes of certain Spots in their Surfaces, and the Noting how long it is e'er those Spots come round again. The Earth's Period is known from the apparent Motion of the Fixed Stars from any Meridian to the same again; which is a Periodical Revolution, or Day. The Moon's diurnal Period is known from the Periodical Month, which is exactly equal thereto; otherwise the same side of the Moon would not be always turned to our Earth, as it certainly is; and that to such a degree of Nicety, that the Menstrual and Diurnal Motions of the Moon have not in the least gain'd nor lost upon each other from the earliest Times of Observation: which is a thing exceeding remarkable, and what will be taken particular Notice of hereafter.

Corollary. By comparing this and the fourth Table together, we learn the true Horary Diurnal

nal Velocity of the Equatoreal Parts of the several Planets: which therefore are as follows.

The *Sun*'s Equator		4.000
Jupiter's	moves in	25.000
Mars's	the space	500
The *Earth*'s	of an	1.030
Venus's	Hour ab^t.	1.600
The *Moon*'s		10

Miles

(9.) The *Quantity of Light and Heat* deriv'd from the Sun to every Primary Planet of this System, and to each of the three Comets, when at their mean distance, has been already set down, according to the Numbers following.

Bodies close by the Sun as	45000
Mercury as	6
Venus as	2
The *Earth* and *Moon* as	1
Mars as	$\frac{4}{10}$
Jupiter as	$\frac{1}{27}$
Saturn as	$\frac{1}{90}$
1st Comet as	$\frac{1}{1108}$
2d Comet	$\frac{1}{2691}$
3d Comet	$\frac{1}{20000}$

These Numbers are easily found; being as the Squares of the Planets distances from the Sun reciprocally.

(10.) The *Eccentricities* of the Orbits of the several Planets, and of the 3 Comets, are thus set down already; supposing each of their middle Distances to be represented by 1000.

In

of RELIGION.

	In Proportion.	In Miles.
Saturn's	55	42,735.000
Jupiter's	48	20,352.000
Mars's	93	11,439.000
The Earth's	17	1,377.000
Venus's	7	0,413.000
Mercury's	210	6,720.000
The Moon's	55	0,013.000

N. B. The *Eccentricities* of the Orbits of the three Comets, are nearly equal to their middle Distances themselves; which are already set down, *pag.* 51. *prius.*

These *Eccentricities* are known, as to the Planets, by the difference of their Apparent Diameters, and so of their *Real Distances*, reciprocally proportional to them in the Aphelia and Perihelia of the Planets; the half of which difference is equal to this *Eccentricity*. But in the Comets, whose farthest Distance is invisible, 'tis known by the meer Subtraction of their nearest from their mean Distance; and noting the difference, this difference is the Eccentricity it self.

The Times in which the several Primary Planets would fall to the Sun, and the Secondaries to their Primaries, if their Projectile Velocites were stopp'd, and they were permitted to fall directly to those Centers, by the Power of Gravity, are by me elsewhere thus stated.

F *Mercury*

		D.	H.
Mercury		15	13
Venus		39	17
The Earth or Moon	would	64	10
Mars	fall to	121	00
Jupiter	the Sun	767	00
Saturn	in	190	00
Nearest Comet		13000	
Middlemost		23000	
Outmost		66000	

See Math. Philos. *Prop.* 23. *p.* 172, 173.

The *Circumjovials*.

The Innermost	would	0	7
The Second	fall to	0	15
The Third	Jupiter	1	6
The Fourth	in	2	23

The *Circumsaturnals*.

The Innermost	would	0	8
The Second	fall to	0	12
The Third	*Saturn*	0	19
The Fourth	in	2	20
The Fifth		14	1
The Moon would fall to the Earth in		4	20

A Stone would fall to the Earth's Center, if there were a hollow Passage, in 21′. 9″.

The way of discovering these **Numbers** is this: It has been demonstrated in the Place already referr'd to, that half the Period of every Planet, when it is diminished in the Sesquialteral Proportion of the Number 1 to the Number 2,

A MAP of the MOON Fig. VI
Pag. 67.

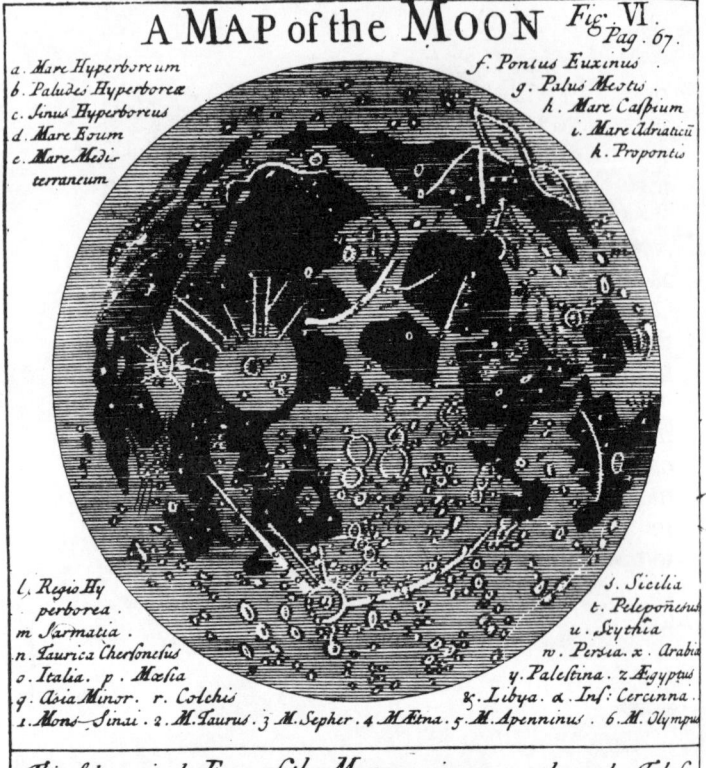

a. Mare Hyperboreum
b. Paludes Hyperboreæ
c. Sinus Hyperboreus
d. Mare Eoum
e. Mare Mediterraneum

f. Pontus Euxinus
g. Palus Mæotis
h. Mare Caspium
i. Mare Adriaticū
k. Propontis

l. Regio Hyperborea
m. Sarmatia
n. Taurica Chersonesus
o. Italia. p. Mœsia
q. Asia Minor. r. Colchis

s. Sicilia
t. Peleponesus
u. Scythia
w. Persea. x. Arabia
y. Palestina. z. Ægyptus
&. Libya. a. Ins: Cercinna

1. Mons Sinai. 2. M. Taurus. 3. M. Sepher. 4. M. Ætna. 5. M. Apenninus. 6. M. Olympus

This Scheme is the Face of the Moon as it appears through a Telescope at the Full, and as described by Hevelius; the Dark Parts are the Sea; the Bright Parts Land, and the long white streaks, the Illuminated tops of Ridges of High Mountains.

The Spots here Described are for the main the same that are continually expos'd to our sight; on Account of the exact adjustment of the Moons Diurnal and Menstrual Revolutions, whereby almost the very same Face is continually turned towards our Earth.

I say nothing of the Librations discovered in its Motions by Hevelius, which make the Parts sometimes hidden to appear to us and thereby afford its Bordering Inhabitants (if such there be) the glorious view of our Earth, which the more remote Ones can never enjoy without Traveling a great way for so uncomon a Prospect.

That the Moon has an Atmosphere about it we have lately discovered; but that Atmosphere being very thin and only visible in Total Eclipses of the Sun, it was not necessary to represent it here.

I. Senex sculp.

or nearly in the proportion of 1000 to 2828, is the Time of its falling to the Center: From which Demonstration it is easy to derive the foregoing Numbers.

(11.) The Moon has *Day* and *Night*, *Summer* and *Winter*, *Mountains* and *Valleys*, *Land* and *Sea*; as also an *Air* or *Atmosphere*, with Clouds and *Vapours*, and a *Moon*, and all after the same manner, in general, that our Earth has them.

See the Map of the Moon, Fig. VI.

That the *Moon* has *Day* and *Night*, is evident from the constant falling of the Sun's Light upon one Hemisphere of the Moon, and the removal of that Light, from *East* to *West*, quite round it, in a Synodical Month, and is visible to our Eyes; which Space is therefore equal to an intire Νυχθημερον, which is 29 d. 12 h. 44ʹ. long by our Computations.

That the Moon has *Summer* and *Winter*, is evident from the Librations of its Body, *North* and *South*, which imply that its Axis is about $6\frac{1}{2}$ Degrees distant from that of the Ecliptick, as our *Summer* and *Winter* is made by the Declination of the Earth's Axis $23\frac{1}{2}$ Degrees from the same Axis. Only it hence follows, that in the Moon, tho' the Day be near 30 times as long as ours, and the Year only equal to ours, In duration; yet that Year is with much less inequality of Seasons, of Heat in Summer, and Cold in Winter, than ours; on account of the much smaller Declination of the Moon's Axis, than of that of our Earth, as compar'd with the Axis of the Ecliptick.

That the Moon has *Mountains* and *Valleys*, every Body that has seen its Face through a Telescope cannot but know; these Inequalities of its Surface, especially near the Limits of Light and Darkness, excepting the Full Moon, being to all most obvious and sensible; as are also a great Number of circular Cavities in other Places, into which the Sun may be perceiv'd to shine, and cast a Shadow, (excepting near the Full Moon) as evidently as the Moon does so here with us. Only it must be noted, that those who have measur'd the Height of the Lunal Mountains, which we stand here very conveniently to do, find them much higher, at least in proportion to its Semi-diameter, if not also in reality, than those of our Earth.

That the Moon has *Land* and *Sea*, or some Parts full of Inequalities, like our Land, which strongly reflect Light; and others smooth and plain, like our Seas, which reflect it more weakly, is, I think, now very clear, not only from the obvious distinction which even the naked Eye makes between the rougher brighter Parts, and the smoother Spots, and which the Telescope does more fully confirm; but more particularly from Mr. *Derham*'s Noble Observation, which I am inform'd was first made by *Hevelius*; that when the Limit of Light and Darkness passes over the brighter Parts, 'tis plainly jagged and uneven; but strait and even when it passes over the darker; which seems to me entirely to determine this Matter.

Astro Theol. Pref. p. 51.

That the Moon has an *Air* or *Atmosphere* encompassing it round, is now, I think, very plain also, from its Appearance in Total Eclipses of the Sun, the fittest Times of all for its Observation, and especially from the two last Total Eclipses, in 1706, and 1715, of which last I have given a full *Account* to the World; and whch

Fig VII

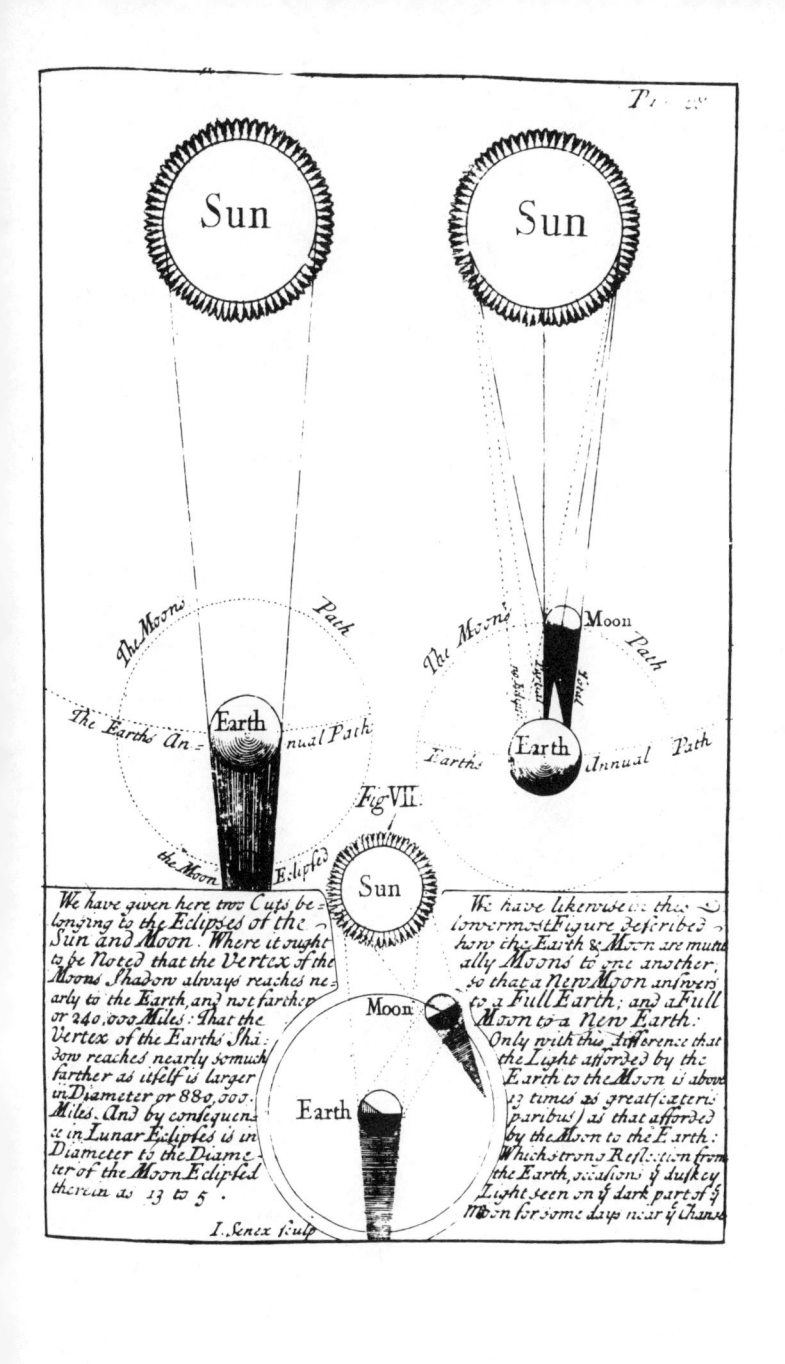

We have given here two Cuts belonging to the Eclipses of the Sun and Moon. Where it ought to be Noted that the Vertex of the Moons Shadow always reaches nearly to the Earth, and not farther or 240,000 Miles: That the Vertex of the Earths Shadow reaches nearly so much farther as itself is larger in Diameter or 880,000 Miles. And by consequence in Lunar Eclipses is in Diameter to the Diameter of the Moon Eclipsed therein as 13 to 5.

We have likewise in the lowermost Figure described how the Earth & Moon are mutually Moons to one another, so that a New Moon answers to a Full Earth; and a Full Moon to a New Earth. Only with this difference that the Light afforded by the Earth to the Moon is above 13 times as great (cæteris paribus) as that afforded by the Moon to the Earth: Which strong Reflection from the Earth, occasions a duskey Light seen on ye dark part of ye Moon for some days near ye change.

J. Senex Sculp.

of RELIGION. 69

which *Atmosphere* will, I hope, be more nicely taken notice of and confirmed at the next Total Eclipse, *May* 11. 1724, which I have prepar'd the Reader for in the same *Account*. Only we must observe, that it appears by the Phænomena, that the Moon's Atmosphere is, in proportion at least, much *higher* than ours, and at the same time much *rarer* also.

That the Moon has *Clouds* and *Vapours*, is evident from the foregoing Assertions, as to its Seas and Atmosphere; and is distinctly proved by the Phænomena of the last Total Eclipse, in the Scheme or Account before-mentioned. Only it must thence be noted, that the rareness of its Atmosphere is so great, that it will not support such gross opake Masses as our Clouds are here; but that the Vapours there rise and fall in a more easy and insensible manner, than is the Case in our Earth, at least since the Deluge of *Noah*: For I am still of the same mind as to the Antediluvian State, which I proposed in my *New Theory*, that the Earth's Atmosphere was [2d *Edit.* p. 245, 247.] at that time comparatively regular and clear, and did then resemble what we find to be the agreeable condition of the Lunar Atmosphere at this Day; I mean by the Rising of the Vapours in the Day, and their Falling down in the Night, in the Form of a gentle Mist; without any of those Opake Masses which we call Clouds; and without any of those violent Storms of Wind, Rain, Thunder, and Lightning, &c which we, to our Sorrow, experience in our Atmosphere since that time.

That our Earth is a *Moon* to the Moon it self, and that a glorious and most useful One also, is not only evident from the Consideration

See Fig. VII.

on of the Astronomical System of the Earth and Moon, but is directly visible to our selves also; it being clear that that Secondary Light which we commonly see in the dark part of its Body, for several Days before and after the New Moon, is no other than Light reflected to the Moon from our Earth, and thence reflected back to us again. Nor is this very surprizing neither; since the Square of the Earth's Diameter is more than Thirteen times as great as that of the Moon; and by Consequence the Light of the Earth is at the Moon in the same proportion greater, than that of the Moon at the Earth. Only it must be here noted, that how considerable soever the Light of a full Moon be to us, and the Light of a full Earth to the Moon, yet that both of them are very inconsiderable, if compar'd with that of the Sun to either of them; as being *cæteris paribus* in the proportion only of the Square of the Earth's Semi-diameter, to the Square of the distance between the Earth and Moon, or, which is the same thing, as the enlightned Hemispherical Surface of the Moon or Earth, whence the Sun's Light is reflectal, to that Hemispherical Surface whose Radius is the distance of the Moon from the Earth, over which that reflected Light is spread: *i. e.* that the Sun's whole Light is about 48000 times as great as that of a Full Moon to us; and about 3600 times as great as that of a Full Earth to the Moon.

N. B. Perhaps therefore the Reason why we have never been able to procure any sensible Heat by burning Glasses, when expos'd to the Moon, is not from any real want of Heat in them,

them, but only becaufe thofe Glaffes have never been large enough to gather Lunar Rays fufficient for that Purpofe.

N. B. The other Secondary Planets, I mean thofe about *Jupiter* and *Saturn*, are too fmall and too remote from us, to afford us fuch Indications of their State, as we have of the State of our Secondary Planet, the Moon. Nor indeed do the primary ones themfelves afford us enow of them to determine in particular their own State, as to many fuch Matters. Only *Jupiter* affords us the Appearance of Belts, or movable Girdles, and befides them, his Satellits afford us that famous Phænomenon of the Velocity of the Rays of Light: And *Saturn*, befides his Five Planets, affords us fuch a Ring encompaffing his Body, as feems to be the moft fingular and curious Spectacle in the whole Syftem. Of thefe Three Phænomena therefore I fhall give fome farther Account, before I proceed to the Comets and fixed Stars.

(12.) As to the Belts, or movable Girdles of *Jupiter*, they feem to be formed by its Clouds, which appear to lie and to move regularly, parallel to its Equator, much after that manner that our Clouds do between the Tropicks; where the conftant Trade-Winds blowing ftill from *Eaft* to *Weft*, muft, in a lefs Degree, caufe our Earth's Surface to appear at a great Diftance with fuch Belts alfo. Nor is it very ftrange, that *Jupiter*'s Clouds lie more copioufly and regularly in fuch a Parallel Situation, than ours do, if we remember the vaftly greater Magnitude of *Jupiter*, than of our Earth; and its much quicker diurnal Revolution

See Fig. VIII.

tion at the same Time. But as to *Jupiter*'s, and other bright fixed Spots, compar'd with the rest of the duller Parts, I take them, as in the Moon, to be Indications of the Diversity of Land and Sea in those Planets; although their much greater Distance from us, makes it hitherto a great deal harder to discover the same more distinctly.

(13.) That the Rays of Light are not a meer and absolute Instantaneous Pressure through a Fluid, but a real Succession of the small Particles of Light gradually, and in Time flowing from the *Sun*'s Body, is proved by the Eclipses of *Jupiter*'s Planets. These Eclipses ever *anticipate* the even Calculation, when our Eye, by the annual Motion, *meets* the Rays of Light reflected from them, whether at their last Egress from the Sun's Light into *Jupiter*'s Shadow, or at their first Ingress into the same Light afterward; and these Eclipses ever come *too slow* for the same even Calculation, when we are going from those Rays; and this still in that Proportion, which implies that the Rays go no less than 81,000,000 Miles, or from the Sun to the Earth, in half a Quarter of an Hour

See Fig. IX.

(14.) That *Saturn* has a broad Ring about itself, like a broad Tin Horizon about a Globe, is now well known, since the Days of the famous Mounsier *Huygens*, who first discovered what it was: And every body that views *Saturn* through a good *Telescope* of Ten or more Feet, may see it very plainly at this Day. This Ring is vastly large; and when measur'd by the Micrometer, and compar'd with *Saturn*'s own

See Fig. X.

Fronting Page. 72

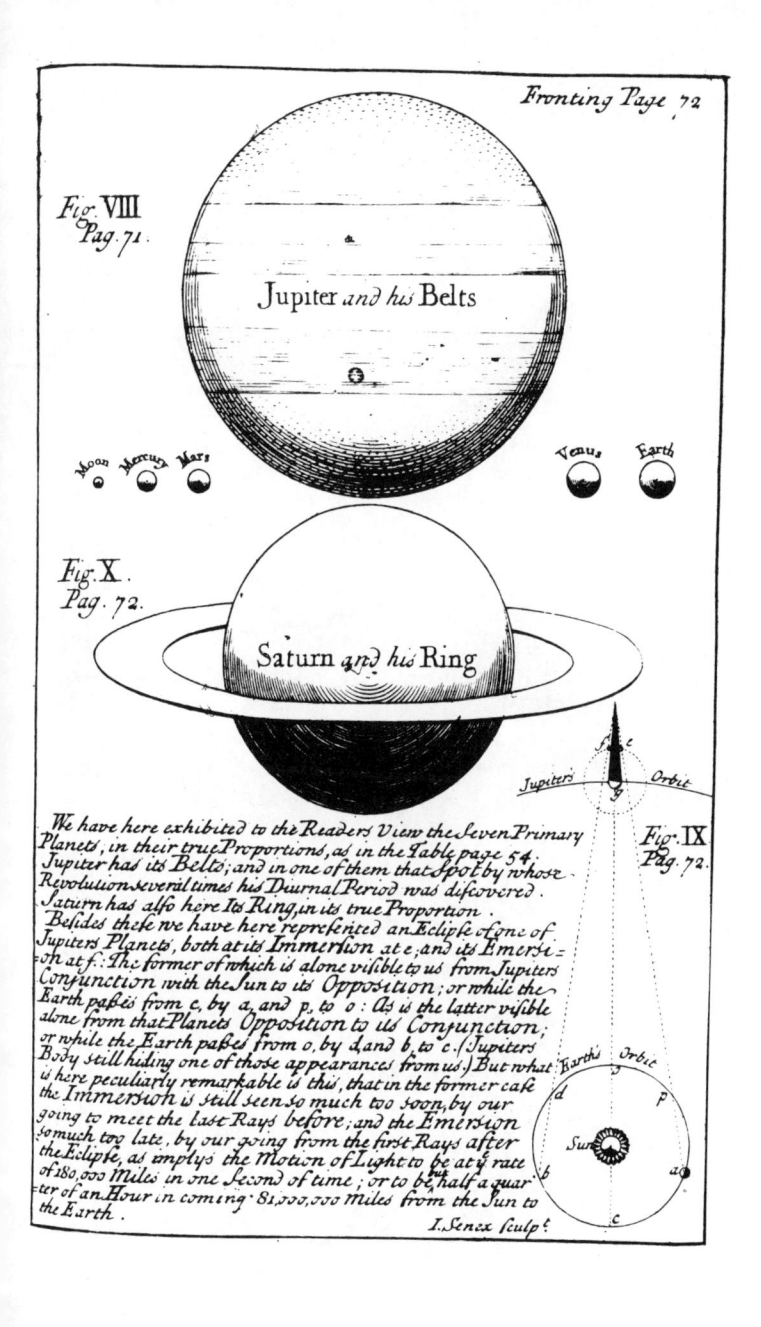

Fig. VIII
Pag. 71.

Jupiter *and his* Belts

Moon Mercury Mars   Venus Earth

Fig. X.
Pag. 72.

Saturn *and his* Ring

Fig. IX.
Pag. 72.

We have here exhibited to the Readers View the Seven Primary Planets, in their true Proportions, as in the Table page 54. Jupiter has its Belts; and in one of them that Spot by whose Revolutions several times his Diurnal Period was discovered. Saturn has also here Its Ring, in its true Proportion.
 Besides these we have here represented an Eclipse of one of Jupiters Planets, both at its Immersion at e; and its Emersion at f: The former of which is alone visible to us from Jupiters Conjunction with the Sun to its Opposition; or while the Earth passes from c, by a, and p, to o: As is the latter visible alone from that Planets Opposition to its Conjunction; or while the Earth passes from o, by d and b, to c. (Jupiters Body still hiding one of those appearances from us.) But what is here peculiarly remarkable is this, that in the former case the Immersion is still seen so much too soon, by our going to meet the last Rays before; and the Emersion so much too late, by our going from the first Rays after the Eclipse, as implys the Motion of Light to be at y rate of 180,000 Miles in one Second of time; or to be half a quarter of an Hour in coming 81,000,000 Miles from the Sun to the Earth.

I. Senex Sculp.ᵗ

own Body, appears to have the Dimensions following.

Inward Semi-Diameter about	55000	
Outward Semi-Diameter	76000	
Difference, or Breadth	21000	
Distance from *Saturn*'s Body on every Side	21000	Miles.
Thickness not exactly known; perhaps	750	

N. B. This Ring casts a mighty Shadow upon large Regions of *Saturn*'s Body, which removes from one Part to another, and causes great Diversity as to Light and Darkness thereon: And, what is probably to them more strange, they hardly know what it is that causes those Varieties: For though we stand conveniently to see what it is, and know it to be a mighty Ring encompassing *Saturn*'s Body; yet is it not so easy for any in *Saturn* to discover it. They must naturally imagine it to be in the Heavens above them; and have, as it seems, no way, but by Astronomical Observations and Parallax, to find out what it is: Just as we find out the Distance and Motions of the Moon, or of any other Heavenly Body. Whether this Ring, or indeed *Saturn* it self, revolve about its Axis, we are not yet assur'd: Nor are we certain whether this Ring be Solid or Fluid. When once our Glasses can shew us any permanent Spots in the Body and Ring of *Saturn*, we shall be able to determine whether there be such a diurnal Motion or not; but till that Time we must be content with Conjectures: For which, if there be room, I should certainly suppose the Ring

Ring to be Solid, and both to have such a Motion; nor only from its Conveniency for any Creatures that may be thereon, but from parity of Reason, and the general Case of the rest of the Heavenly Bodies: Not one of which are yet known to be destitute of such a Motion; although some of them have not yet afforded us an Opportunity of certainly determining the same.

See the Account of Comets towards the end of my Mathematick Philosophy, at large.
Ibid. p. 415.

(15.) The Number of the Comets is very considerable; to be sure much greater than of the visible Planets. Our famous Dr. *Halley*, by searching into the Histories of them, and comparing the Observations made about them, has given us a Catalogue or Table of so many as he could find well enough describ'd to afford Foundation for determining their Obits; which Table I have elsewhere given the Reader.

This Number is 24, all which have appear'd within the last 400 Years, in these Parts of the World. He has also observ'd, that out of these 24, Three had their Orbits and Circumstances so very like, and the Intervals of appearing so nearly equal; and that Two others had their Orbits and Circumstances so very like also, that he justly concluded it exceeding probable that the former Three were one and the same; and in some Measure so, that the latter Two were also one and the same Comet; returning the first after 75, the latter after 129 Years. Sir *Isaac Newton* also discover'd, and in the new Edition of his *Principia*, Published his Discovery, that the last most eminent Comet of 1680, 1681, towards the end of its Appearance, bent its Course so much inward from a Para-

P. 464, 465.

Parabolick Line, as to shew its real Trajectory to be *Elliptical*; and this in such Proportion, that its Period of returning must be more than 500 Years. After which Discovery I my self, for several Reasons, supposing this to be the same Comet that caused the Deluge, did accordingly guess the Period to be either 575, or 504 Years; according as it had made either Seven or Eight Revolutions since that Time; and drew up Tables upon both those Hypotheses, when the same Comet must have appear'd afterwards, in order to search whether they did so or not; but not having either *Hevelius*'s or *Luvienetz*'s Histories of Comets then by me, I could not immediately confirm my Hypothesis any further. But in a little time I found, that Sir *Isaac Newton*, and Dr. *Halley*, had compleated what I wanted; and had discovered that just such a Comet had appear'd the 44th Year before the Christian *Æra*, which was the Year that *Julius Cæsar* was slain; as also, Anno Dom. 531, or 532; and again A. D. 1106; and lastly, A. D. 1680, 1681, and this still, after the forementioned Interval of about 575 Years; and that they accordingly did justly conclude it to be the very same Comet that appeared in those several Years. So that we have the Orbits of only 21 Comets; and the Periods at the most of only Three of them yet kown.

N. B. The first of these Three appear'd probably 1304, and 1456, but more certainly 1531, 1607, and 1682; and as there is little reason to doubt, will by Consequence appear again 1758, and 1832, and so every 75 Years afterwards. The Second of them appear'd A. D. 1532, and not improbably again 1661, and so it may be expected

expected to return 1789, and again 1918, and so every 129 Years afterwards. The Third having last appear'd 1680, 1681, and having its Period no less than 575 Years, cannot return till *A. D.* 2255. But then, as to the rest of the Comets, we cannot yet foretel the Periods of their Revolution, for want of Ancient exact Histories and Observations, but must leave their Determination to future Ages; which will, no doubt, if Astronomical Learning continues in the World, be in Time fully discovered, and known by our Posterity.

See Fig. XI. *N. B.* The Observations shew that the Comets are about the Bigness of the Planets, and have vast Amospheres about them; the Central Parts thereof being denser than the Superficial; as also, that the Vapours and Clouds thereof are hurried about in great Disorder and Confusion, like Planets in a Chaotick State: It also appears from the like Observations when near the Sun, that their Atmospheres wind themselves round like a revolving Globe, and ascend towards the Regions opposite to the Sun; as if the rarest Vapors whereof they are compos'd, were carried up by the Solar Rays; as also that those Tails are generally the longest which arise from the greatest Nearness to the Sun.

N. B. The Motion of Comets being some from *North* to *South*, others from *South* to *North*; some from *East* to *West*, others from *West* to *East*, in all Plains and Directions; and this from the very Sun downwards, to the Regions vastly beyond *Saturn* upwards; and that Motion appearing most exactly regular, without the least Retardation by any sensible Resistance in their several Courses, it is thence most certain, that
there

This is the Representation of the lowest part of the Tail of a Comet, near its Perihelion, with the purer part of its Atmosphere winding itself into its Tail, And the Cloudy part of the same placed round about the Central Solid As it Appear'd to Dr. Hook through a Telescope.

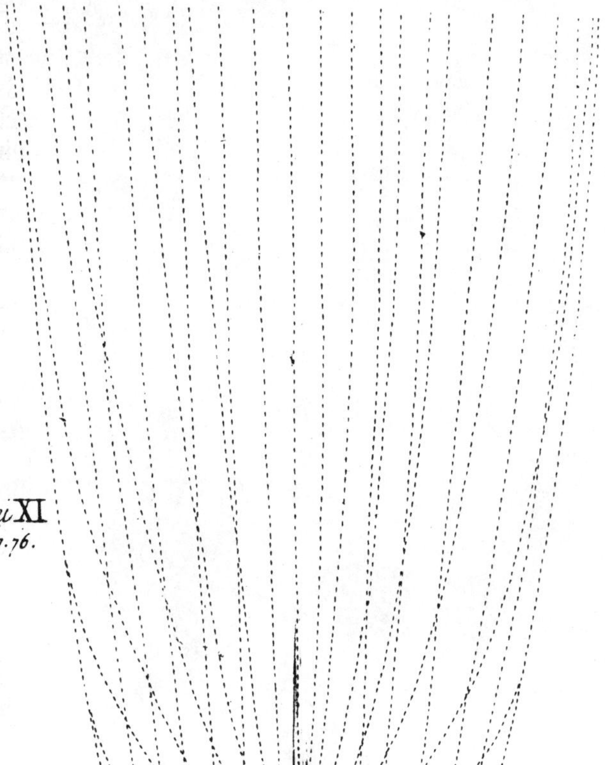

Figure XI
Pag. 76.

I. Senex sculp.t

there are no *solid Orbs*, as the Ancients supposed; and that there is no *subtile Matter*, as the *Cartesians* imagin'd; but that all the vast Spaces between, and beyond the Planetary System, are an immense Void or Vacuity, as to sensible resisting Matter; and admit ordinarily of nothing but of the Rays of Light, unless it be near the Comets with their Atmospheres and Tails; and near the Planets with their Atmospheres; all which Particles put together are almost nothing in comparison of those vastly prodigious, those immensely numerous Vacuities which are interspers'd between them, as we have already seen from an exact Calculation, *Pag.* 19.

(16.) The Fixed Stars, visible to the naked Eye of the Acutest Observers, are in Number considerably under 2000, and those seen only through Telescopes, about 10 or 20 times that Number. Their Distance is found by Mr. *Huygens*'s conjectural Method to be about 2,200,000,000,000. And the Distance of some, from Dr. *Hook*'s and Mr. *Flamsted*'s Annual Parallax, which I look on as much more certain, about the third Part of that distance, or 700,000,000,000 Miles. However, 'tis certain that this Distance is vastly great if compar'd with that of the Planets and Comets, when they are remotest from the Sun; and that in particular the famous Comet of 575 Years Period, which goes sometimes about 14 times as far off the Sun as *Saturn, i. e.* about 11,200,000,000 Miles, is not then near enough to them to be altered by that of the Fixed Stars Gravitation towards, or Influences from them. These Stars seem to be of the

See my Astron. Lect. III, IV.

the same Nature with the Sun, as shining with their own Native Light; and continuing fixed in the Centers of their several Planetary and Cometary Systems, as the Sun does.

(17.) The Fixed Stars seem to be really of very different Magnitudes, yet not of such very different Distances from this System as is now generally supposed. As to the first Assertion, it seems very agreeable to the rest of the visible Bodies in the World, whether Planets or Comets, which are of very different Magnitudes also; and seems confirm'd from the second Assertion; for if the Fixed Stars visible to the naked Eye be at no very different Distances from our System, that vast apparent Inequality of Light which they send hither, and according to which they are ranked under six or seven different Magnitudes on our Globes and Planispheres, will be next to a Demonstration, that they are themselves really of very different Magnitudes also. Nor in that Case will it be proper to place the Telescopick Stars at any vastly farther Distances, since they do not at all seem different from the other, only still gradually smaller. Now that the Fixed Stars, visible to the naked Eye, are not at any very different Distances from our System, is most probable, because the best Method we yet have of knowing those Distances, I mean their Parallax, determine that the Distance of those Three which have been try'd, tho' all of different apparent Magnitudes, is very nearly the same. I do not deny that some Fixed Stars may be vastly farther off than others; and that there may be Systems of Worlds scattered every where in the Universal Immense Void: But I say, that if farther

ther Observations confirm this Parallax, and any sort of Equality thereof, we must accommodate our Opinions to our Evidence, and in that Case must suppose, that the grand Systems themselves of Sun, Planets, Comets, and Fixed Stars are like the Parts of such a System, vastly remote from each other; nay, perhaps, out of the reach of each others Discovery also. But as to this noble Theory, we cannot be at all positive till the Parallax and Distance of the Fixed Stars be more nicely observ'd, and the Astronomical World better satisfy'd about it; the doing ofdwhich I would therefore earnestly recommen to the publick Consideration.

(18.) Several of these Fixed Stars, especially of the smaller sort, do sometimes disappear, and new ones appear; and some of them do appear or disappear, look Brighter or Duller by Turns; and this sometimes after certain Intervals of Time also. This is a known Fact; and has in some measure been noted from almost the earliest Ages of Astronomy. But then, what should be the Cause of such mutable Appearances among these Fixed Stars, is by no means yet discovered; nor have I hitherto ventur'd to propose any Conjecture about them. However, since others have already begun their Hypotheses, which seem to me commonly either intirely precarious or absolutely impossible, I shall make bold here to offer my own, which shall not only be free from such strong Objections, but agreeable to the nearest parallel Case of the World. We know that the Sun it self, the only Fixed Star, as I may call it, that is sufficiently within the reach of our Telescopes, has several Times not a few *Maculæ* or *Spots* upon his Body; which frequently

quently become *Faculæ*, or Parts brighter than the rest; and which come and go by turns. How many there may possibly be of these Spots at certain times we cannot say; but this is plain, that we do not know but so many of them may sometimes arise, as may, in good part, cover over the Sun's Surface, and render its Light and Heat very Weak and Dull. These Spots may be again dissipated, and become *Faculæ*, or Brighter than ordinary. These *Maculæ* and *Faculæ* may sometimes, by Turns, gain ground on one another, after certain Intervals of Time, and cause the Sun to grow Darker and Brighter periodically. Nay, we do not know, but these *Maculæ* may sometimes, especially in Case the Sun were smaller, cover over the greatest Part of its Surface, so as to extinguish, or at least to obscure its Light; which yet in Length of Time may be overcome, and the Sun may recover its former Splendor, if not one greater than that before. Since therefore such Phænomena of the Fixed Stars are like what our Sun appears by known Observations to be in some sort liable to, I think it the best Guide, as to what we see to happen in others of them; and that the *Maculæ* and *Faculæ* of those Stars may cause these surprizing Appearances. Nor can I easily think in any other manner, about those six Spaces of Light, or Starry Mists, which have been lately discovered in the Heavens, than by Analogy to what we know of Things of a like Nature; *viz.* that they are a Company of very small Fixed Stars, as invisible to us with our ordinary Telescopes, as the known Telescopick Stars in the Milky Way are to our natural Eyes, which give such an irregular Appearance of indistinct Light also.

Transact. Philos. N°. 347 *for* A.D.1716.

PART

Part IV.

Certain Observations *drawn from the foregoing* System.

(1.) SINCE Matter is entirely a Paffive Subftance; no fpontaneous Motion or Action, even in Brutes, can be derived from it, much lefs can that Active and Free Being, the Soul of Man, be juftly fuppofed to be material.

(2.) Since Bodies once exifting will continue to exift, and that for ever, in the fame State of Reft, or uniform Motion, along ftrait Lines, wherein they once are, at leaft with the alone Concurrence of the Firft Caufe; the Projectile Motions of the Heavenly Bodies require no new, or particular conftant Acts of Power for their continuation in that State. But,

(3.) Since none of the Heavenly Bodies move uniformly in ftrait Lines, but all of them rather unequally, and all in Curves, they are every one impell'd, and that perpetually, by fome External Power, Efficacy, Force, or Influence; and thereby

thereby obliged to revolve in such Curves; which Power we have already proved to be that of Gravity.

(4.) If that Power of Gravity were suspended, all the whole System would immediately dissolve; and each of the Heavenly Bodies would be crumbled into Dust; the single Atoms commencing their several Motions in such several strait Lins, according to which the projectile Motion chanc'd to be at the Instant when that Influence was suspended or withdrawn.

(5.) Since pendulous Bodies receive no sensible Resistance in their Internal Parts; and since both the Planets and Comets move prodigiously swift with the utmost Freedom, and without any sensible Resistance through the Æthereal Regions, 'tis certain there is no *Subtile Matter* pervading the Universe, as some have supposed.

(6.) All the Solutions, therefore, of the Phænomena of Nature, which depend on the Supposition of that *Subtile Matter* are intirely false, and contrary to the plain State of our *System*.

(7.) To suppose a *Plenum*, or that the Universe is entirely full of such Subtile Matter, is utterly vain and ungrounded; nay, contrary to the most certain Observations.

(8.) Since the particular Proportion ever obtaining as to the Power of Gravity, I mean that of the Duplicate of the Nearness of Bodies, is not any necessary Result from the Nature of Matter, or any Laws of Motion in the World; it is plain that this Proportion is no way owing to any Mechanical Cause or Necessity whatsoever, but intirely to free Choice, Prudence and Judgment. (9.) Since

(9.) Since all Bodies are equally capable of Rest, and of all Degrees of Velocity whatsoever; but are in their own Nature no way determin'd to any of them; that nice Adjustment there is of the projectile Velocity to the Attractive Power through the whole Universe; whereby the Planets both primary and secondary revolve nearly in Circles, and the Comets nearly in Parabola's, is no way owing to any Mechanical Cause, or Necessity whatsoever; but entirely to free Choice, Prudence, and Judgment.

(10.) Since all Bodies are equally capable of being originally impell'd every way, and of having their projectile Motions in any Direction whatsoever; and since all the Planets, both Primary and Secondary, have their projectile Motions almost perpendicular to the Lines from their Central Bodies, which was absolutely necessary to their Motion at nearly the same Distance from the Sun; this nice Adjustment of the Direction of their projectile Velocity, whereby they became fit for the Habitation of Animals, and without which they would have been almost useless in the World, is no way owing to any Mechanical Cause, or Necessity whatsoever; but entirely to free Choice, Prudence, and Judgment.

(11.) Since all Bodies are equally capable of Rest as of Motion, and that in any Time, and with any Velocity; and since there is no original Connection between the projectile Directions and Velocities, of any two or more Bodies which now revolve about the common Centers of their own Gravities; without which Revolutions about those Centers, the present System of the Universe

Universe could not be supported; and yet without the most exact Adjustment of those Directions and Velocities to one another, the Directions parallel, but contrary to each other, and the Velocities in a Proportion reciprocal to those Bodies themselves, such a Revolution could not be performed: Such an Amazing and Mathematical, and Universal, Adjustment of these Circumstances, cannot be owing to any Mechanical Cause or Necessity whatsoever, but must arise entirely from free Choice, Prudence, and Judgment.

(12.) Since all Bodies are equally capable of possessing any part of Space whatsoever; and since the Planets, both Primary and Secondary, and the Comets, with their several Degrees of Velocity, might be placed at any Distance from their Central Bodies; that nice Adjustment there is of their several Distances from those Central Bodies, to their several Velocities; whereby the Planets all revolve nearly in Circles, and the Comets nearly in Parabola's, can be no way owing to any Mechanical Cause or Necessity whatsoever; but entirely to free Choice, Prudence, and Judgment.

(13.) Since all Bodies are equally capable of any Direction whatsoever, and yet all the Planets, both Primary and Secondary, do revolve about their Central Bodies, according to one Direction almost from *West* to *East*; this particular Direction of the Annual Motions of all the Planets is no way owing to any Mechanical Cause or Necessity whatsoever; but entirely to free Choice, Prudence, and Judgment.

(14.) Since

of RELIGION. 85

(14.) Since all Spherical Bodies are equally capable of turning round upon any Axis, or according to any Direction whatsoever; and yet the Sun and all the Planets, whose Diurnal Motions are already discovered, do revolve about their own Axes nearly from *West* to *East*; this particular Direction is no way owing to any Mechanical Cause, or Necessity whatsoever; but entirely to free Choice, Prudence, and Judgment.

(15.) Since all Spheres revolving about their own Axes are equally capable of turning round with any Velocity, and in any Period; and yet all the Heavenly Bodies that do so revolve, keep within proper Limits, agreeably to the State of every such Sphere; as particularly we find to be the Cases of our Earth, and of *Mars* and *Venus* our Neighbouring Planets: This due Proportion of the Diurnal Motion is no way owing to any Mechanical Cause, or Necessity whatsoever; but entirely to free Choice, Prudence, and Judgment.

(16.) Since all Bodies revolving about another, are equally capable of moving in any Plains whatsoever; and yet the several Primary Planets move almost in the same Plain about the Sun, and the Secondaries all in their same several respective Plains about their Primaries, this exact Direction of the Planets into the same Plains is no way owing to any Mechanical Cause, or Necessity whatsoever; but entirely to free Choice, Prudence, and Judgment.

(17.) Since the greatest, as well as least Planets are equally capable of being plac'd near, as well as far off the Sun, the placing the largest Primary Planets the most remote, and the least nearest the Sun, whereby the several Motions

G 3 continue

continue most undisturb'd, and many fatal Consequences, which would otherwise happen, are prevented; is no way owing to any Mechanical Cause or Necessity whatsoever; but entirely to free Choice, Prudence, and Judgment.

(18.) Since the Secondary Planets are equally capable of revolving about their Primaries at the smallest, as at any other Distance whatsoever; and yet they all of them are situate so far off, as to cause no dangerous Tides in the Primaries Ocean, which a much greater Nearness would certainly have done; this due Place of the Secondaries from their Primaries, is no way owing to any Mechanical Cause or Necessity whatsoever; but entirely to free Choice, Prudence, and Judgment.

(19.) Since all the Planets are equally capable of revolving round the Sun at the smallest, as greatest, Distances, and yet they are Situate in a Mean, so as neither to be scorch'd with its Heat, nor frozen up with Cold, for want of it; as we particularly find to be the Case of our Earth: This proper Situation of the Planets, is no way owing to any Mechanical Cause or Necessity whatsoever; but entirely to free Choice, Prudence, and Judgment.

(20.) Since the Comets do revolve in very oblong Ellipses, quite through the Planetary Regions; and this in vastly long Periods, according to all manner of Directions, and in all Situations of their Plains, contrary to the Laws every where observ'd among the Planets; 'tis very evident that the Intentions and Designs for which they are fitted, are very different, at least in their present State, from those for which the other are accommodated.

(21.) Since

(21.) Since that Immechanical Power of Gravity, which is constantly exercis'd in the World, is proportionable to the Quantity of each Body to which it belongs; which Quantity is vastly unequal in the several Celestial Bodies; it is thence certain, that the Author of that Power must be a Being that exactly knows, and takes perpetual Notice of all those Bodies whatsoever, in all the Variety of their Parts, and Magnitudes.

(22.) Since that Immechanical Power of Gravity which is constantly exercis'd in the World, is not of one even and constant Quantity, but vastly unequal, according to the Squares of the different Distances of the Bodies affected with it; it is thence also certain, that the Author of that Power must be a Being that exactly knows, and takes perpetual Notice of the Distances of all those Bodies whatsoever, in all the Variety of their Parts and Magnitudes.

(23.) Since all the Effects of this Immechanical Power of Gravity do constantly obtain, and all the Consequences of that Power are ever found true in Fact, throughout the Universe; (abating only the Case of Miracles, not here to be consider'd) it is certain, that the Author of that Power can and does move all Bodies, how great soever, and with what Degree soever of Velocity, according to that due and fixed Proportion; without the least Opposition or Contradiction, either from the Matter to be moved, or from any other Agent whomsoever.

(24.) Since the Spherical Figures, with the Original Native Light of the Fixed Stars, and all other Circumstances, do shew that they belong to our Universe, or grand System, and are subject to the same Law of Gravity which our particular System is governed by; it follows, that the foregoing Consequences concerning the Author of that Law already drawn, as to one System, are also true, relating to all the others.

(25.) Since all the Motions in our Solar System must be so far at least retarded, as their Passage through a Medium every-where penetrated with the Rays of Light must imply; which Retardation, how small soever it be in itself, must in sufficient Length of Time become sensible, (as it begins to be already in the Case of the Lunar Period,) it follows, that the several Parts of this System do by Gravitation naturally and constantly, unless a miraculous Power interpose to hinder it, approach nearer and nearer to the Center of Gravity of this System; and in a sufficient Number of Years will actually meet in the same Center, to its utter Destruction.

(26.) Since this entire grand System of Things is subject to this Power of Gravity; and since that Power of Gravity has its Effects as well among the Fixed Stars, with their several Systems, as in our Planetary and Cometary World, about the Sun; and since withal, the Sun and Fixed Stars do not revolve about one another, or about any common Center of Gravity, as the Planets and Comets do; which Motion alone, according to Mechanical Laws, can hinder the Effect of that Power of Gravity: it follows, that the several Systems, with their several Fixed Stars or Suns, do naturally and constantly, unless a Miraculous Power interposes

of RELIGION. 89

fes to hinder it, approach nearer and nearer to the common Center of all their Gravity; and that in a fufficient Number of Years, they will actually meet in the fame common Center, to the utter Destruction of the whole Univerfe.

(27.) Since Power can be exerted no where but where the Being which exerts that Power is actually prefent; and fince it is certain, as has been fhewn, that this Power is constantly exerted all over the Univerfe, 'tis certain that the Author of the Power of Gravity is prefent at all Times in all Places of the Univerfe alfo.

(28.) Since this Power has been demonstrated to be Immechanical, and beyond the Abilities of all Material Agents; 'tis certain that the Author of this Power is an Immaterial or Spiritual Being, prefent in, and penetrating the whole Univerfe

(29.) Since the Sun and Fixed Stars fend out perpetually, and with the utmoft Velocity, Rays or Corpufcles of Light and Heat from themfelves; and fince we fee with our Eyes that there is not any fuch Equality of thofe Stars on every Side, as might induce us to believe there can be an equal Circulation of thofe Rays from one Syftem to another; and fince we find by Mutations in our Sun, and by the Parallel Mutations in feveral of the Fixed Stars, that thefe very Suns themfelves, the Fountains and grand Supports of the feveral Syftems, are equally liable to Decay with the reft of the Univerfe; 'tis hence alfo plain, that all thefe Suns and Syftems are not of Permanent and Eternal Conftitutions; but that, unlefs a miraculous

lous Power interposes, they must all, in length of Time, decay and perish, and be rendred utterly incapable of those noble Uses for which at present they are so wonderfully adapted.

N. B. Although the External Parts of the Heavenly Bodies, with their Nature and Uses, may be most *easily* and certainly determin'd from Fact and Observation, yet does there not want Arguments whereby we may come at the Internal Parts or Regions of the same, their Nature and Uses, at least from very probable Considerations. Thus we know from Observation in Comets, that there are large Central Solids inclosed in their Atmospheres, indissoluble by the utmost Heat in their greatest Nearness to the Sun: We also know from the like Observations, that Comets are about the Bigness of Planets, and that the Atmospheres of Comets do best answer the Chaotick or Primary State of Planets, of all other Bodies in the Universe. *See at the End of my Mathematical Philosophy.* We know farther by Demonstration, that if there be any Central Cavities within such Solids, the Effects of the Power of Gravity will be there so equipois'd on every Side, that there will appear to be little or nothing of such a Power at all: And, lastly, we may know, in some Measure, by Observations and Demonstrations compar'd together, whether there be such Cavities in them or not; and in which of the Heavenly Bodies they are the most considerable, as we shall see presently. *See Math. Philos. Prop. 44.* This general Observation thus Premis'd, I come to the Fifth Part of this Treatise, to give my Conjectures as to the several Natures and Uses of all the Parts of this System.

PART.

of RELIGION.

Part V.

Probable Conjectures *as to the Nature and Uses of all the Parts of the* System *of the visible World.*

(1.) THE Sun and Fixed Stars are, to be sure, on their external Regions or superficial Parts, most intense Fire or Light; and the grand Fountains of that Fire and Light which is in the whole visible Universe, and without which there could be no such thing as a *Visible Universe*, or Useful System at all. So that there can be no doubt of the general Nature and Use of those external Regions. Nor perhaps shall we be far out of the way, if we suppose those Parts of the Sun to be 10000 Planets or Comets all on Fire.

(2.) The Planets, both Primary and Secondary, appear, as to their visible external Regions, or superficial Parts or Atmospheres, to be like to that Planet we live upon, the Earth; or most convenient and well contrived Habitations for all sorts of Sea and Land, visible and gross Animals;

Animals; with such Plants as are useful for any of their Preservation and Sustenance, during their continuance thereon.

(3.) The Air expanded about the several Planets, which, as to their Elastical Parts, are corporeal, but invisible, appear to be the proper Places for the Habitation of not wholly Incorporeal, but Invisible Beings; or of such as have Bodies made of too subtle and aerial a Texture and Constitution to be ordinarily seen by our Eyes, or felt by our Hands. And if it be considered, that while all the ancient Prophane Traditions, and Historical Accounts, as well as the Sacred Writings, which assure us of the Existence of such invisible Beings about our Earth, do at the same time assure us of their inhabitating in our Air, which is the only apparent Place, according to the best Philosophy, where such invisible Beings, not destitute of all Bodies, can possibly inhabit; it will justly deserve our Consideration, whether this be not the noblest Design and Use of our Air; tho' at the same time its lowest Regions be an *Atmosphere* also; or be useful in Respiration, in Refraction, &c. and so fitted as to elevate and let fall the Vapours belonging to our Earth, for the Support of the Creatures, in grosser Bodies, inhabiting thereon.

See Prop. after my Boyle's Lectures p. 287--297. or Serm. & Eff. p.170, --178. Meteor, p.68--72.

(4.) The external Regions of Comets, which by passing through such immense Heat when nearest, and such prodigious Cold when farthest off the Sun; and by the confused and Chaotick State of their Atmospheres, do evidently appear incapable of affording convenient Habitations for any Beings that have Bodies, or Corporeal Vehicles, whether visible or invisible to us; seem rather

of RELIGION.

rather fitted to cause the grand Mutations of Nature in the Planetary World; by bringing on Deluges in their Descent, and *Conflagrations* in their Ascent from the Sun; as I have elsewhere more fully Discoursed. New Theory, 2d Edit. p.437, 438. and p.440, &c.

(5.) These Comets, with their Atmospheres and Tails, seem also fitted, as to their external Regions, to be a very uneasy, hot, and fiery Habitation when near, and a very uneasy, cold, and chill one, when far off the Sun, and this both in their Surfaces, and in their Airs.

(6.) As to the Internal Parts or Regions of the Sun, Planets, and Comets, they seem to be Concave, and to include vast open Spaces within. This Conjecture which is no way contrary to any other Phænomena of Nature, I ground particularly upon the small Inequality there is in Fact between the Polar and Equatoreal Semidiameters of those Heavenly Bodies which have diurnal Revolutions about their own Axes, compared with the much greater Inequality there would naturally be between them, if they had not such Central Cavities. For Example, If the internal Parts of the Earth were of the same Density with the External, it is Demonstrated by Sir *Isaac Newton*, that the Polar Semi-diameter or Axis would be about 17 Miles shorter than the Semi-diameter of the Equator. If the Central Parts were much Denser than the rest, (as on all mechanical Accounts they ought certainly to be) these 17 Miles would be mightily increas'd, and probably amount to some Hundreds. Yet is that Semi-diameter in Fact but about 31 Miles shorter than the other. Whence it is probable, that the greater Density in the deeper Regions, is compensated by the lesser Density, Princip. 2d Edit. p. 382--- 387.

Denſity, or rather total Cavity of the Central Regions themſelves. And this Reaſoning is ſtill vaſtly ſtronger in *Jupiter*; where the difference of Semi-diameters ought to be much more remarkable, and to amount to ſome thouſands of Miles, while yet it is therein but juſt ſenſible; and that only by the Uſe of the beſt Inſtruments and Obſervations we have in Aſtronomy.

(7.) It is not improbable therefore, that thoſe Central Cavities may be ſo fitted by Providence, as to afford Habitations to ſome Creatures, as well as the external Surfaces, the Land, the Water, and the Air, have appear'd to do; tho' this in different Circumſtances, as to the different Bodies, the Sun, the Planets, and the Comets. See Dr. *Halley*'s Conjecture to ſomewhat the like Purpoſe, *Tranſact. Philoſ.* N°. 347. *for A. D.* 1716.

(8) If the Sun has ſuch a Cavity for Habitation, it muſt be fenced from the Heat of his more external Parts by a vaſtly thick Wall or Partition; which that there may be even of many thouſand Miles, the prodigious largeneſs of its Diameter, and the little comparative depth of the penetration of Heat through ſolid Earth, do Demonſtrate. Nor would ſuch a central Cavity in the Sun be in danger of any pernicious Heat, tho' it were large enough to hold, on its inward Surface, as many Creatures as the external Surfaces of all the Planets and Comets put together could contain; as he that conſiders the Tables, *pag.* 54. and 56. before will eaſily believe. Perhaps ſuch a Degree of Heat may be deriv'd from the outward to the inward Regions, as will ſuit ſome of the Purpoſes of the Great Author

See pag. 54.
 rius.

of RELIGION.

Author of Nature therein: But then all its Light muſt be deriv'd from ſome other Cauſe, than from the outward Parts of the Sun.

(9.) If the Planets, or any of them, have ſuch Cavities for Habitation; becauſe they keep nearly at the ſame Diſtance from the Sun, in the ſeveral Parts of their Courſe, their Heat as well as Light, muſt come moſt probably from within alſo, as not being ever Recruited in their Revolutions.

(10.) If the Comets, while they continue ſuch, have the like Cavities for Habitation; Part of their Heat, becauſe of their Acceſs to the Sun every Period, may be derived from it, and recruited every Revolution: But then all their Light, as well as that of the reſt, muſt be deriv'd from within alſo.

(11.) That the Earth in particular has ſuch a Cavity, ſeems clear from Scripture, as well as it may be conjectured from Aſtronomy. For when many, at leaſt, of the Souls departed out of the World, are there repreſented as *gone down* into the inviſible World; as *deſcended* into the Place beneath, or as *gone down quick into the Pit*; and when our Bleſſed Saviour is there ſtill repreſented as upon his Death, *going down into the Inviſible World, and deſcending into the lower Parts of the Earth*; nothing ſeems ſo agreeable both to Nature and Revelation, as this Hypotheſis; which ſuppoſes ſuch a Receptacle for Inviſible Beings beneath, as exactly anſwers to the foregoing Deſcriptions.

See Gen. xxxvii. 35. Num. xvi. 30, 33. Ezek. xxvi. 20. xxxi. 16. Rom. x. 7. Eph. iv. 9. 10.

(12. If there be any ſuch Cavities and Receptacles for living Creatures, and the Things neceſſary for their Suſtenance, in the Central Regions

Regions of the Sun, or of the Planets, or Comets, 'tis certain their State and Circumstances must be very different from those on the Surfaces of the Planets. They must all live in Concave Spheres, which must hinder all Intercourse betwen them and this visible World: Nor can they have any Philosophical Evidence that there is such an External World at all; which is the Case of the rest of this Universe, as to us, if we, with all the visible Stars, Comets and Planets, be our selves included in such a Cavity; which is not absolutely impossible to be suppos'd. But then, as to the particular Circumstances of such Creatures, their way of Living, and the Course of Nature and Providence, and Divine Revelation relating to them, I shall not venture here to propose any particular Conjectures about them; only hinting this, that the Power of Gravity from the External Parts being in this Case none at all, as we have elsewhere observ'd, there may be therein such a World as is that we here see, with the like Sun, Planets, and Comets; only that they must be so much less in Quantity and Largeness, as the greater Narrowness of their Cavities requires: Yet still such as the Imagination will not be able to distinguish from our larger visible Universe it self.

See Math. Philos. Prop. 44.

N. B. I hope that all judicious Persons will distinguish what I venture barely to *Conjecture* sometimes in this, from what I usually *Assert* in the other Parts of this Treatise: It being ever proper, if any one proposes Conjectures to the World, which are often of considerable Advantage, as they afford Hints for Enquiry further, and Occasions for the Discovery of Truth, to distin-

distinguish them still from Assertions; which ought generally to be built on considerable Evidence, before they are proposed as true to the World : Which Distinction between *Assertions* and *Conjectures* I always aim to make, and always beg of my Readers, that in their Perusal of my Papers, they will ever make the same also.

PART VI.

Important Principles *of* Natural Religion, *Demonstrated from the foregoing* Observations.

BY the *Observations* already made from the true System of the World, it appears,

(1.) That the Souls of even all Brute Creatures are *Immaterial, i. e.* not compos'd of that dull, unactive, insensible, solid, passive Substance, which we call *Matter*, or *Body*, and of which all the Visible and Sensible World about us is compos'd. For it certainly appears, that all this Substance, which we properly call *Material Substance*, or *Body*, is so far from a Capacity of Sensation, Thought, Activity, and Moving it self and the Body, which are the known Properties of the Souls of even Brute Creatures themselves, that it is not capable of any active Property at all; being ever mov'd, impell'd, attracted, and directed entirely as other Bodies or Powers act upon it, and not otherwise. This I say is the natural
Conse-

Consequence of all that Physical Knowledge of Matter or Body, which Observation and Demonstration leads us to. We can Philosophically trace material Impulse, and the material Images of External Objects, in some Measure, even in Brutes, through the Organs of Sense, up to the Brain, or the Fountain of Sensation and Action. But there Mechanical Causes end, and Material Effects cease; there that Agent or Substantial Being, which Sees, and Hears, and Feels, and Tastes, and Smells, and Joys, and Grieves, and Directs, and Moves, and Remembers, and Hopes, and Fears, is present, and directly comes to be considered by us. But then it comes to be consider'd entirely as an Invisible or Immaterial Agent, or Substance, different from the Eyes, and Ears, and Hands, or Feet, and Palate, and Nostrils, and Nerves, and Brain, and Blood, and Animal Spirits of the Brute Creature; just as we consider the Organist that guides and directs the Pipes, and Sounds, and Stops of his Organ, as entirely different from those Pipes, and Sounds, and Stops themselves. Not that I pretend to tell of what particular Kind or Sort of Substance, that Sensitive and Active Being is compos'd, either in Brutes, or in Men, or in any Superior, Invisible, and Intellectual Beings. That Substance may be different in every different Kind. But that we do still observe from their Phænomena, that these Souls are different in their Properties and Actions, in almost all their Properties and Actions, not only from their own Gross Bodies, but in general from what we call Matter or Body, from all that we call Matter or Body in the Universe. Immaterial they are indeed, and

Different from the Material Body, according to all the Notions and Experiments we ever have of these Brutes, and of Matter. But that does not directly inform us *what* they are, because we know not what *Substance* is in general, nor what the particular Substances of any particular Material or Immaterial Being are. But by comparing the Properties of Matter, with those of the Souls of even Brutes themselves, we plainly and evidently perceive that those Souls are not Matter. 'Tis true, we cannot determine how far the Destruction of the gross Body of a Brute affects the Soul, or sensitive Agent thereto united: We do not know whether those Souls are Immortal or not: We do not know whether they utterly perish with their Bodies; or whether they only continue in a kind of inactive insensible State, till they actuate other Bodies again or no, and so perpetually. And the Reason is plain, we have no way of knowing such Secrets of Nature and Providence, about Invisible Substances, by Observation, or Experiment: And the Author of Nature has not been pleas'd to discover such Things to us, by Revelation. If I were to *Guess* where I cannot *Know*, I should imagine that all such Immaterial Souls do ever, by their peculiar Nature, perceive and feel whatever is present to them, and no more; and that they endeavour to act in a way suitable to the Impressions and Passions thereupon arising in them, and no farther: That therefore, when they are out of a Body, their Sensations are so few and narrow, and their Powers so small, that they are in a sort of Sleep, Silence and Inactivity; that when they are in Bodies, (which are most wonderful Structures and Machines, contriv'd triv'd at once to convey the Impressions of most numerous, and even most distant Objects to them, and to follow their Directions and Actions thereupon,) while those Bodies keep in Order, those Souls are alive, vigorous, and active; but when the Bodies are dissolv'd, or their Contexture destroy'd, their Souls return to their former State of Sleep, Silence, and Inactivity; though without a real Annihilation; and so without any Incapacity of Revival and Reactivity: I mean upon Supposition, that the Author of Nature, or any of those Ministers by whom he Governs the World, affords it another Opportunity for such a Revival and Reactivity again. This is, I say, what I should *Guess*, as to the State of the Immaterial Souls of Brutes, if I would indulge my self in such Uncertainties. And whether the *Psalmist* does not favour such a *Conjecture*, where he speaks thus, *When God hideth his Face* from such Creatures, *they are troubled: When he taketh away their Breath they die, and return to their Dust: When He sendeth forth his Spirit they are Created, and He reneweth the Face of the Earth:* I shall leave to the Reader's Consideration. But as to the Proposition it self; I mean, that what perceives and acts even in Brutes, is a Being or Agent properly and entirely distinct from that Body in which it acts, and is truly a Being or Agent Immaterial. This I take to be the natural Result of Philosophick Reasoning, from Fact and Experience; and by no means to be set aside, because we are not able to solve all Difficulties thereto relating: Which we are rarely, or indeed never capable of doing, in the last Resort, in any part of Knowledge whatsoever. But as

Psal. civ. 29, 30.

to those who, to avoid all such Difficulties, pretend that Brutes have no Souls, no Sensation, no Action of their own, and are meerly Corporeal Clock-work, and Machines; they yield me this Point, that if they have Souls, and do really Perceive and Act themselves, those Souls, those Principles of Sensation and Action, must be Immechanical, and Immaterial: But then, they pretend to disbelieve that, which seems to me almost as plainly matter of Fact and Observation, as any of Mr. *Boyle*'s Experiments whatsoever. And I care not to answer such an extravagant Objection, which it is next to impossible to suppose, that any sober Proposer can, in earnest, believe himself in his own Proposal of it. And when the Scripture assures us, that a *Righteous* or *Merciful Man regardeth the Life of his Beast*; and this in Opposition to the *Cruelty*, which the *tender Mercies of the Wicked* or *Savage Men* are affirm'd to have; it most naturally implies, that those *Beasts* are themselves really sensible Creatures, and not incapable of feeling the Effects of the Care, or of the Cruelty of their Masters towards them.

Prov. xii. 10.

(2.) We hence learn more certainly, that the Souls of Men are *Immaterial*. For if we have found that the lower Faculties and Operations of the Souls of Brutes, require us to allow them to be Immaterial, how much more must we do so as to Human Souls? For if we proceed higher to the Rational, the Intelligent, the Penetrating, the Free, the Active, the Sagacious Soul which is in Man; if we consider its vast Capacities and Faculties; and that it can, and sometimes does, act contrary to all the material Impressions or Temptations which the sensitive Soul lays before

before it; contrary to all worldly Views and Motives, all corporeal and terrestrial Interests, and that it can and sometimes does freely chuse Poverty, Misery, and Persecution in this World, out of regard to God and Religion, and the Happiness of a Future State; This Rational Soul, I say, must needs be of so vastly higher a Nature, of such vastly nobler Faculties, of such a vastly superior Rank in the Creation, that He who can once suppose, that there is nothing in this Case but such Matter as we have Knowledge of in the World, with its Accidents, is in a fair way to believe any Properties may belong to any Beings, and that there is really no Distinction between a Square and a Circle; no Difference between Strength of Reason, and the Sound of an Organ; no Preference between the Arguments used in a Theological Dispute, and the Collission of Elastical Bodies in Motion; between a piece of Clock-work manag'd by Wheels and Springs, and an Human Soul govern'd by Reason and Religion; which Confusion of Things, entirely different in their own Nature, seems to me so absurd and preposterous, so wild and aukward, that I have not Patience to set about a more operose Confutation of it. Those who are under any Temptation to believe such Notions, which seems to me no less foolish than the wildest Dreams of Ignorance and Superstition themselves, may consult * Dr. *Clark*'s excellent Confutation of a late Writer upon this Head. Tho', after all, I can hardly think that Writer weak enough to have been in earnest; and if he were, I should have thought that too much Honour was done him and his crude Notions, when they were vouchsafed the favour of so masterly

* See *his Letter to Mr. Dodwell, and its four Defences.*

ly a Confutation. But this is too like a Digression to be farther infisted on in this Place. However, we may obferve here, how agreeable this Immateriality of Human Souls is to the Sacred History of their firſt Original; where after the *Lord God had formed Man out of the duſt of the Ground*, the material Body, perhaps with its ſenſitive Soul alſo; He diſtinctly, from above, infus'd the Rational: *He breathed into his Noſtrils the Breath of Life, and Man became a living Soul.*

Gen. ii. 7.

(3) We hence learn not only the *Immateriality*, but the *Immortality* alſo of Human Souls; or that the Deſtruction and Diſſolution of the Body, with its Senſations, will not deſtroy or diſſolve the Rational Soul united thereto; but will leave it ſtill capable, not only of exiſting, but of acting in another State; if it pleaſe God ſo to diſpoſe of it, that it may have proper Opportunities for doing ſo. So far, as I take it, true Philoſophy carries us here; I mean, it obliges us to put ſuch a difference between the rational Soul, and the brute Body, that the Ruin of the one will no way infer the Ruin of the other; and that therefore, ſince Divine Revelation aſſures us of the living and acting of the Soul in the intermediate State, and alſo of its Return to the Body, and acting therein again after the Reſurrection, this is all agreeable to ſound Reaſon and Philoſophy, to good Senſe and the Laws of Nature: Tho' ſtill all this, without aſſerting ſuch a neceſſary Immortality, or Eternal Duration in Happineſs or Miſery, as is independent on the Power, and Will, and Laws of the Author of Nature; to which all the Enjoyments, and Faculties, and Perceptions of a Human Soul,

may

may still be owing hereafter, altho' the Substance it self of that Soul should, of it self, when once created, continue to exist, as all real Beings seem to do, without any particular Interposition of Providence for such their Continuance. Philosophy, Mathematical and Experimental Philosophy, obliges us to suppose, that the Soul will continue to exist after Death, and will therefore be still *capable* of Action and Enjoyment, of Happiness and Misery. Divine Revelation assures us this separate Soul *shall* Act and Enjoy, shall partake of Happiness or Misery in a lower State and inferior Degree before, and in an higher State and superior Degree after the Resurrection: So that Reason and Religion supply, and support, and confirm each other, and, upon the whole, assure us of the Truth of this grand Principle of all Religion, especially of the Christian, That this Life is not the only, or the principal Stage on which we are to Act; that this World is not the only or the principal Time for our Happiness, or Misery; but that, after this frail and mortal Life is ended, which is only a short State of Tryal and Probation, we must live a longer one of Enjoyment hereafter. Which Truth is of that Importance for us to be satisfy'd in, that nothing of either Natural or Supernatural Knowledge, which tends thereto, ought to be neglected by us. Nor may we here omit the exact Agreement of this Natural Truth of the *Immortality* of Human Souls with Divine Revelation, particularly with our Saviour's own important Words upon this Head: *Fear not them which kill the Body, but are not able to kill the Soul: but rather fear him, who is able to destroy both Soul and Body in Hell.* Mat. x. 28.

(4.) We

(4.) We hence learn the *Being of God*, the first Intelligent Cause and Author, the just Owner and Possessor, the Supreme Lord and Governor, the constant Preserver and Disposer of all Things. This Foundation of all Religion, the Belief of a Supreme Deity, is the first, the most natural and obvious Deduction of Human Reason, even from the Contemplation of the most common and ordinary Appearances of Nature; from the Growth of every Plant, and the Succession of every Season, and the general View of every Heavenly Body, and every Creature about us. And there have certainly been no Nations or People, of the usual Capacities of Mankind, but have ever drawn this Consequence in all Ages of the World. So that if this Inference be not the *Voice of Nature* it self, we shall be at a great loss to find other Truths, requiring any Reasoning at all, that can deserve to be so stiled. And no wonder, since the Argument is the very same by which, from the Contemplation of a Building, we infer a Builder; and from the Elegancy and Usefulness of each Part, we gather he was a skilful Architect; or by which from the View of a Piece of Clockwork, we conclude the Being of the Clockmaker; and from the many regular Motions therein, we believe that he was a curious Artificer. Which Deductions he who is not *Able* to make, has not the *Reason*; and he that will not allow them to be *Just*, has not the *Honesty* of the meanest Countreyman. 'Tis true, that if this sort of arguing were confin'd to Childhood, or Folly; to the Age of Ignorance, or the Temper of Ideots; if the more nicely we viewed the World, the less Reason we found

found to admire its Contrivance; and when we were come to the top of Enquiry and Examination, we lost all the Occasions of our Wonder and Adoration: If, I say, this were the Case, a sober Person might think fit to suspend his Assent, and to cast about for some other Solution of the Phænomena of Nature. But in case the wise and examining Man still finds vastly stronger and more numerous Reasons for the Acknowledgment of the Divine Existence, than the Fool, or the careless Enquirer does; so that if he spends his whole Life in the pursuit of this sort of Knowledge, he perceives new Arguments every where crowd upon him to the same purpose; which is the known Case, as to Experimental Philosophy, at this Day; He who is still resolv'd to suspend his Assent, and either to wrap himself up in wilful Scepticism, as if he knew nothing; or to try how far he can be absurd enough to believe, that the World is it self the only God, the only Eternal, Omnipotent, All-wise Being; or, which is yet more absurd, that all the Wonders in its Contrivance came meerly by Chance and Accident; and will continue by Chance and Accident, till by the like Chance and Accident they all come to nothing again: He, I say, who acts thus, does certainly, if ever Man did, *dare operam ut cum ratione infaniat*; takes great pains to shew himself, with great Learning, the most egregious Fool in the World. While true Wisdom or Philosophy would teach him to assent to the Apostle *Paul*, when he justly affirms, That *the Invisible Things of God are clearly seen from the Creation of the World; being understood by the things that are made; even his Eternal Power and Godhead*. So that Rom. i. 20.

Men

Men are without Excuse for Atheism. But this general Reasoning is so obvious, and so common, that I shall not here enlarge upon it; but rather apply my self to Demonstrate the particular Attributes and Operations of God from the particular Phænomena of the World already set down; as being a thing less common, and of greater Advantage. Accordingly from what has been before advanc'd, we learn,

(5.) That the World has not been from all Eternity, but that God was the *Creator* of it, and that He, and He alone, at first Disposed and Ordered the several Parts of the Universe, into that wise and wonderful Structure in which we now see them. I do not here mean to intermeddle with that more intricate Problem of the proper *Creation* of the *Matter of the Universe out of Nothing*; because the Phænomena of Nature give us no Indications either way: nor, as I understand it, does Divine Revelation ever directly concern it self with it. Only, that I may not be mistaken, I declare my own Opinion to be still, as it has ever been, against the Eternity of Matter, and for its Original Creation out of nothing by the Almighty Power of God. But then, I am not only of *Opinion*, but am fully *Satisfy'd* from the plain Phænomena of Nature, that the World was, in the more ordinary Sense, originally Created, and at first put into that State in which, for the main, we still find it by the Divine Power, Wisdom and Goodness. I have already shewed, that the present System of Things, acting according to those Laws of Motion and Nature which are now fixed in the World, cannot possibly have been *a parte ante*, and cannot possibly be *a parte post* Eternal: Much less is it possible, that one

Pag. 88, 89, 90, prius.

of Religion.

one little Corner of it, such as our Earth, should have been, or should be hereafter Eternal by it self. This pretended Eternity of the World, is indeed so far from the Result of any just Reasoning, or Philosophical Evidence, (and to any Divine Revelation it never yet, that I know of, made the least Claim;) that I dare appeal to the entire System of Nature, whether there appear one single Argument for, or Indication of such an Eternity, either *a parte ante*, or *a parte post*, in the whole Universe. I profess, I know none: And unless Men be so weak as to leave Fact, Nature, Experiments, and Mathematicks, for the Subtilties of Metaphysicks, and the Cobwebs of Abstract Notions, they must believe the World not to have been, nor to be, Eternal. But for those that are subtil enough to deny the Reality of Motion, or the Freedom of Human Actions, because they are not able to account Metaphysically for Motions Migration from one Subject to another, as they speak, or for the Mode and Seat of Human Freedom in the Soul; while their own Senses and Observations do every Day of their Lives assure them of the Reality of them both; they are Persons fit for the Atheists Purpose in this Matter, and will be proper Patrons for the Eternity of the World. He that, with *Ocellus Lucanus*, can prove the World to have *no End* in Point of Duration, because it is of a round Figure, which certainly in another Sense has *no End*; or He that can demonstrate from abstract Reasoning about the Nature of Matter, and that Equality of Motion he may suppose to be in every Part of the Universe; that a certain Clock or Watch will of it self go for ever, tho' at the same time he sees such wearing of the Wheels and Pivots, such decay of the Spring, and such Rust and Foulness over the whole, (besides the Necessity of its being wound up every Revolution) as must, by Calculation, put a Stop to its Motion in 20 Years time; Such as these, I say, are also Persons rightly disposed for the same Doctrine of the World's Eternity. But for the rest, I mean for all the sober Part of Mankind, who are govern'd by common Reason and Experience, they will, I believe, yield to the Facts and Arguments already used; and because 'tis thence certain, that the present World is not, cannot be naturally Eternal, will readily ascribe its Origin to the great Author of Nature, to God himself: They will easily allow, that the *Belief in God the Father Almighty, as the Maker of Heaven and Earth*, is no less a certain Doctrine of Natural Religion, than it is a primary undoubted Article of the Christian Faith: Nor will they wonder, that the *Jewish* Legislator begins his *Archæology* with this Assertion, *In the Beginning God created the Heaven and the Earth*; as a very proper, a very necessary, and certain Preliminary to that Divine Legislation he was to lay down, in vertue of the Authority of the same great Creator.

Gen. I. 1.

(6.) We learn from the foregoing System of the World, that God, the Creator of it, is an *Eternal* Being, and has existed before all Time, and all Worlds, even absolutely from everlasting. This Attribute of God, of a proper *Eternity*, or *Necessity of Existence*, without any Cause or Beginning whatsoever; (for these Properties seem to me to infer each other, if not to be one and the same;) appears to be one of the hardest Notions that

that a Human Mind can take in. Yet is it the moſt certain of all others. For 'tis moſt apparently plain, that if the Firſt Cauſe and Original Being does now, or ever did exiſt, as we have ſhew'd he now does, and ever did ſince the World began; He muſt have exiſted from all Eternity, otherwiſe He muſt have at firſt *produced* himſelf *into Being* when He *was not*; or muſt *Act* and *Create* before He *was*; which all the World owns to be abſolutely impoſſible. This Attribute, we ſee, is not deriv'd from any particular Phænomenon of the Univerſe; but from them all, and every one together. Nor do any who believe the foregoing Corollary, concerning the Exiſtence of God, ſo much as pretend to doubt of the *Eternity* of that Exiſtence; ſo that I ſhall not need to enlarge any farther upon this Attribute in this Place. Only we may obſerve, how agreeable this *Eternity* of God, here gathered from the Light of Nature, is to Divine Revelation, which aſſures us, that *Before the Mountains were brought forth, or ever he had formed the Earth or the World, even from Everlaſting to Everlaſting He is God.* Pſal. xc. 2.

(7.) From the foregoing *Syſtem* we learn, that God, the Creator of the World, does also exerciſe a continual *Providence* over it, and does interpoſe his general, immechanical, immediate *Power*, which we call the *Power of Gravity*, as alſo his particular immechanical *Powers of Refraction, of Attraction, and Repulſe*, &c. in the ſeveral particular Caſes of the Phænomena of the World; and without which all this beautiful *Syſtem* would fall to Pieces, and diſſolve into Atoms. On which Occaſion, the *Apoſtolical Conſtitu-*

Constitutions speak as agreeably to Philosophy as to Religion, when they say, *The whole World is held together by the Hand of God.* For, as we have already observ'd, though we should not think it necessary to suppose a particular Interposition of the Supreme Being, in the Conservation of the Natures or Existence of Things, and of their Original Projectile Motions, which once begun may continue of themselves, without any new or particular Support; (in which Case however not a few think it necessary to introduce the same, and that the *Conservation* of Things is no other than a kind of *continual Creation* of them;) tho', I say, we should not think it necessary to suppose, that continual interposing Providence in the material World, in order to its Preservation, yet can we not avoid allowing it in that grand immechanical Power of the Universe, I mean that of Gravity, and in the other immechanical Powers of the same Nature above-mentioned. These do continually act in the World, and alter the Places and Motions of Bodies perpetually; which makes me commonly chuse to call them here *Powers* rather than *Laws*, as the *Power* of Gravity or Refraction, rather than the *Law* of Gravity or Refraction; and so for the rest. For this I take to be the Difference between a *Power* and a *Law*, speaking strictly; that a *Law* belongs to such Rules as necessarily flow from some Property of Bodies, without any new Action exercis'd thereupon; as that *Bodies once in Motion continue for ever to move with that Degree of Velocity which they once have, without Intermission, along those strait Lines, according to which those Motions are already directed*: But a
Power

L. V. 7. init.

of RELIGION.

Power is here *Such a Rule, by which the Bodies are constantly moved out of those Lines, and from those Velocities they naturally had, into other Lines and Velocities distinct from them*; which plainly implies a real Action, a true Force, Impression or Influence actually exerted upon them; whether we call it Impulse, or Attraction, it matters not; and this perpetually also: Which is therefore quite different from the foregoing Case, as to the ordinary Mechanical Laws of Motion; and supposes a real Agent, and He sufficiently Active, and Powerful also to remove all such Bodies through the Universe perpetually. For I desire those who think otherwise, to tell me, how unactive Matter, without such a continual Exercise of Force upon it, can be continually oblig'd to leave its natural, even, rectilinear Motion, and to move faster, or slower, in a curve Line? Can it, of it self, admit of a *Law of Gravity*, and of it self ever exert this *Clinamen*, this *Departure* from its proper Course? But this is to suppose it not *unactive Matter*, but an *intelligent Self-moving Being*, which alone is capable of understanding of Laws, and acting according to them. In short, it seems to me most evident from the Phænomena of the World, that all such Laws or Powers as we are now speaking of, which are many and wonderful, and yet absolutely necessary to the Preservation of the present System, are the real Effects of the continual *Power* and *Providence* of God himself, for the Conservation of the *Universe*; and that whenever he pleases to suspend such Exercise of that Power and Providence, the World it self will dissolve into Atoms, and its present Form

I suffer

suffer an utter Destruction. I conclude this Head in the elegant Words of the *Psalmist*, when he celebrates the great Creator, for some remarkable Instances of his *Providential Care* over his Creatures: *O Lord, how manifold are thy Works! in Wisdom hast thou made them all: The Earth is full of thy Riches. So is this great and wide Sea, wherein are Things creeping innumerable, both small and great Beasts. There go the Ships: There is that Leviathan, whom thou hast made to play therein. These wait all upon thee; that thou may'st give them their Meat in due Season: That thou givest them they gather: Thou openest thine Hand, they are filled with Good. ---The Glory of the Lord shall endure for ever; the Lord shall rejoice in his Works. ---I will sing unto the Lord as long as I live; I will sing Praise unto my God while I have my Being.* But to proceed. We farther learn from this System of the World,

Psal. civ. 24, &c.

(8.) That this Supreme God, the Creator and Preserver of the World, and the Author of the Power of Gravity, and of all other the Immechanical Powers in the Universe, is a *Free Agent*, no way limited by any *Necessity* or *Fate*, but acting still by *Choice*, and according to his own *good Pleasure*. This Attribute of the Divinity, without which the Supreme Being himself would be below Mankind, a meer Fatality, and no way worthy of any Veneration, and Love, and Gratitude from his Creatures, is fully demonstrated under so many of the foregoing *Observations* from the System of the World, that I scarce need quote any one of them in Particular. And I appeal to all that we now know of this entire Universe about us, whether

whether there be one single Indication of such a rigid Fatality, and Necessity therein. 'Tis true, such strange Reasoners as *Hobbs* and *Spinoza*, &c. pretend by Metaphysick Arguments to demonstrate this Fatality and Necessity, even as to the Actions of the Supreme Being, or τὸ πᾶν it self. But then, they do by the like Metaphysick Reasoning, as strongly pretend to demonstrate the same Fatality and Necessity of Human Actions also; which last we are sure, from our own certain, constant Experience, all our Lives long, is utterly false and ridiculous; and therefore we have no Reason to depend on the like Subtilties in the other Case: Especially while instead of Metaphysick Subtilties on one Side, we produce the plainest Experiments, Observations, and Demonstrations from Nature, and the System of the Universe, on the other: And while Dr. *Clarke* has, with great Sagacity, shew'd the Inconsistency of that Metaphysick Reasoning, and the Freedom of the Divine Being, and of Human Actions, even in their own Way, both in his Sermons at Mr. *Boyle*'s Lectures, and Answers to an Anonymous Author already quoted, to which I shall refer those Readers, who have a mind to deal in that way of Reasoning. As for my self, while I clearly see that the constant Experience of all Men, and the entire Phænomena of the whole Universe, directly prove the Freedom of Human and Divine Actions, I am not much concern'd what may be alledg'd to the contrary from Metaphysick Uncertainties, especially when I find it every-where confirm'd by Divine Revelation also; which always eaches us to render such Praise and Thanksgiving to Almighty God, for the Mercies of Creation and Providence, as would be perfectly ridiculous, if all such his Operations were entirely fatal, and involuntary. *Bless the Lord, O my Soul, and all that is within me, bless his Holy Name. Bless the Lord, O my Soul, and forget not all his Benefits: Who forgiveth all thine Iniquities, who healeth all thy Diseases: Who redeemeth thy Life from Destruction, who crowneth thee with loving-Kindness, and tender Mercies. Bless the Lord, all his Works, in all Places of his Dominion: Bless the Lord, O my Soul.*

<small>Psal. ciii. &c</small>

(9.) We learn from the true System of the World, that this Supreme God is an *Intelligent* and *Omniscient* Being, and that he knows the entire State and Condition of every Body contain'd in this entire Universe, at what Distances, and in what Circumstances soever; and that in every Moment of Time from the Beginning of this System, till its Conclusion. This is deriv'd from Observations 21st and 22d foregoing; and is a most unquestionable Deduction from the Phænomena of the Universe. I use the Word *Omniscience* here, as such Words are commonly used in Scripture, and Ancient Authors, for such a Degree of Knowledge, whether in it self absolutely infinite or no, which extends every-where to the whole visible Universe, and takes in all the particular Parts of the same, to the utmost Limits of our Examination and Computations, without Exception. For in this Sense only, can the Phænomena of Created Beings, which must be every way finite, become Demonstrations of the Attributes of their Uncreated Original Creator; which alone can be, strictly speaking, esteemed absolutely

of RELIGION.

absolutely Infinite. Nor are we here, or in other Cases, concerned any farther. For if the Supreme Author of our Universe, does certainly and exactly know whatsoever is, or is done in this Universe, what does it concern us, whether he equally knows whatsoever is, or is done in any other Invisible or Imaginary Systems beyond it? Although he who believes that this Omniscience, or other such infinite Attributes, belong truly to the Supreme God, its Author, to the utmost Extent of this grand System, must be a very strange Person, if he can deny, or doubt whether the Omniscience of the same Supreme Being, extends to the rest of the other invisible Systems also; supposing there be such other Systems, and that they are the Workmanship of his Hands also, as well as ours. All which confirms the Words of *Elihu*, when he thus exhorts the great Example of Patience; *Hearken unto this, O Job, stand still and consider the wondrous Works of God. Dost thou know when God disposed them, and caused the Light of his Cloud to shine? Dost thou know the Balancings of the Clouds, the wondrous Works of him which is perfect in Knowledge?* [Job xxxvii. 14, 15, 15.]

(10.) We learn further from the true System of the World, that this Supreme God does not barely know and understand all the Bodies, and all that is done by all the Bodies, in this entire grand System, but he most *Wisely* and *Prudently*, and *Sagaciously Orders and Disposes of all the said Bodies*, and particular Systems thereto belonging. This is such a known and frequent Consequence of the foregoing *Observations* from the true System of the World, that almost one half of them demonstrate the undoubted

doubted Truth and Certainty of it. Nor could there be any Occasion for enlarging on this Head, were it not that the Moral or Living World, does not here always seem to agree with the Natural or Astronomical one. In the latter there is plainly and every where Marks of such Exactness, Harmony, Prudence, Sagacity, Wisdom, and Conduct, that not only perfectly Convinces, but Amazes and Astonishes us, even all of us, who thoroughly consider the particular Instances, the innumerable, clear, irrefragable Instances of the same, in every Part of the Universe: And he must be stupid to the utmost Degree, who can go a Course of Mechanicks, of Anatomy, of Botanicks, and especially of Astronomy, without the most satisfactory Conviction in this Case. Those natural Sciences, particularly Astronomy, will fully demonstrate what the Sacred Writings have already inform'd us of; not only that *God telleth the Number of the Stars, and calleth them all by their Names*; that *great is our Lord, and of great Power*, and that *of his Understanding there is no Number*; but also, that *His Works are manifold, and in Wisdom has he made them all*. But then, how it comes about that we do by no means perceive the same Exactness and Harmony, in the Moral and Living World, which we every where see in the Material and Physical, is a great and noble Problem; though not belonging to this Place. I have my self occasionally touch'd upon it in my other Writings; but acknowledge it to deserve a much larger and fuller Disquisition. In the mean time, I must confess, that I have met with no Accounts thereto relating, so Authentick, so Rational and

Psal.cxlvii. 4, 5.

civ. 24.

so

so valuable, as those we have in an Ancient, but too much despised Book of Primitive Christianity; I mean the *Recognitions of Clement*; which I sometime since Translated into *English*, for the Advantage of the Unlearned, but Inquisitive Reader; to which therefore I must refer him for his Satisfaction in this Case. Only so far the present Proposition may justly tend to his Satisfaction, as it certainly Demonstrates the Proportion, Harmony, and Decorum every where provided for by the Supreme Being in the Natural World, and by consequence affords us the greatest Reason to believe, what by the Light of Nature we cannot but expect from Him, that in the last resort and upshot of Things, we shall find the same Proportion, Harmony, and Decorum provided for, with regard to all his Living and Rational, that is, to his Principal, which are already so remarkable and surprizing, as to the rest of his Inanimate and Irrational, which are but his Inferior Creatures.

(11.) We learn farther from the true System of the World, that the Supreme God, who Made and Governs it, is a most *Powerful* and *Almighty Being*; whom nothing can resist, and against whom nothing can oppose it self in the whole Creation. This is not only the natural Result of this World's derivation from God, and its receiving all its Powers and Abilities from him; which must needs imply, that the Author of them all is still of greater Power; and that no created Might can oppose it self against that Might by which it was it self created; but is a direct Consequence from the 23d Observation before-going. And that we may have some particular Notion of the Greatness of the Divine Power, and Almighty Efficacy in this our own System, let us consider the Greatness of the Bodies it every where moves; and the Velocity with which it moves them. The former you may find in the Table, *pag.* 56. and the latter in that *pag.* 53. whereby it appears, that the Planets alone, which are continually moved, are together above Four Hundred Millions of Millions of Cubical Miles in Magnitude; that the Velocity wherewith they are mov'd, in their Annual Motion, is, at a mean, about 52000 Miles in an Hour; and that the Velocity of the Corpuscles of Light is still vastly greater, and no less than Six Hundred and Fifty Millions of Miles in the same time; and all this has been so continually from the beginning of the whole System to this very Day; and that without the least proper Resistance or Opposition from either any of these Bodies themselves, or from any other Power or Agent whatsoever. On which account we may every one of us well say with Holy *Job*, after God had made an Eminent Representation to him of his own Omnipotence, and of *Job*'s Weakness; *I know that thou canst do every thing, and that no Thought of thine can be hindred:----Wherefore I abhor my self, and repent in dust and ashes.*

Job xiii. 2, 6.

(12.) We learn farther from this true System, That the Supreme God, who made and governs the World, is every where substantially and really *present* through the whole; or is at all Times, and in all Places, *Omnipresent*. This is a most direct Consequence of the Divine Knowledge, and Wisdom, and Power, the Attributes and Actions of the Supreme Being, continually exerted throughout the whole Universe. Nor

of RELIGION.

Nor can we any more conceive them actually exercis'd *where*, than *when* that Being, whose Attributes and Powers they are, does not it self exist; which last is by all Men allow'd to be grosly absurd and impossible. Nor does either the Sun, or any other Being, afford us any thing, like a Virtual, as distinct from a Substantial Presence and Efficacy. The Sun indeed sends out Rays of Light, which Rays operate where the Sun is not: But then both of them operate only where themselves are. Nor is any other Notion consistent with common Sense, or the possibilities of Things. For to say that a Being Acts *where* it is not, is to say in effect, that *Nothing* Acts in that place; or that the *Effect* produc'd, has in that Place *no Cause* to produce it; which are the grossest Absurdities and Contradictions possible. But then, as to the Vastness of the Extent of this Presence of God, through this grand System, including all the Systems of the Fixed Stars also, it is to us hitherto unlimited and undetermin'd; tho', in all probability, in it self not really Infinite. However, so far we are certainly upon Fact to suppose the Divine Omnipresence to reach, and to be present, as we discover the Effects of the same; I mean so far as the visible Universe extends; which we know, on the lowest Computation, must be nearly that of a Cube of 1,400.000,000.000 Miles Diameter, which contains near 3,000,000,000. 000,000. 000,000,000,000.000, 000.000, *i. e.* three Sextilions, or three Millions of Millions of Millions of Millions of Millions of Millions of Cubical Miles. An amazing Space this,

this, and as to any Power of Imagination, scarcely to be distinguish'd from Infinite Space it self! And so far, to be sure, the Omnipresence of God extends it self. Nor can even these Limits inclose, or limit the Presence of the Supreme Being; who as He is by Demonstration present every where within, and between all the Parts of the several immense Systems, so, no question, is his Presence extended as well beyond the Grand System it self, as we know it is beyond this Particular System wherein we live. But since the vast Visible System of the Universe is that with which we are alone concern'd, and such as even wearies and amazes our Faculties, when we attempt so much as to imagine its Immensity, I shall wade no farther into that unfathomable Abyss of Infinite Extramundane Space; the nicer Consideration of which, like that of Infinite Duration or Eternity, is evidently too large for our finite Thoughts; and does ever more Astonish and Confound, than Profit and Edify Mankind. And no wonder, since 'tis highly probable that both of them, as to their inmost Nature, and largest Extent, are alone knowable by that High and Lofty One, who, in the Prophet's Expression, *Inhabits Eternity*.

Isa. lvii. 15.

I conclude this Head in the apposite Words of the *Psalmist*, of *Jeremiah*, and of the *Apostolical Constitutions*, which shew that Reason and Revelation agree in this important Attribute of the Divine Omnipresence. *If*, says the *Psalmist*, *I ascend up into Heaven thou art there: If I make my Bed in Hades, behold thou art there. If I take the Wings of the Morning, and dwell in the uttermost parts of the Sea; even there shall thy Hand lead me, and thy right Hand shall hold me.*

Pf. cxxxix. 8, 9, 10.

me. And, says God by *Jeremiah, Am I a God at hand, and not afar off? Can any hide himself in secret places that I cannot see him? saith the Lord. Do not I fill Heaven and Earth? saith the Lord.* And say the *Constitutions, Thou art He who art in Heaven; He who art on Earth; He who art in the Sea; He who art in finite Things, thy self unconfin'd by any thing. For of thy Majesty there is no Boundary. For 'tis not ours, O Lord, but the Oracle of thy Servant who said;* And thou shalt know in thine Heart that the Lord thy God He is God, in Heaven above, and on Earth beneath, there is none besides thee.

Jer. xxiii. 23, 24.

L. VII. 35.

Deut. iv. 39.

(13.) We learn farther from this System, that the Supreme God, the Maker and Governor of the Universe, is, in his own Nature and Substance, *Immaterial.* This is also a most direct Consequence from the true System of the World, whereby it has appear'd that God, the only Author of the Power of Gravity, does act, and is present to the inmost Parts of all solid Bodies; nay that he is equally present, and equally acts in those inmost Parts of all solid Bodies, as in any empty Spaces themselves, and withal that this Action upon all Bodies is not like to material and mechanical Impulse, which is ever on the Surface only; but penetrates those Bodies themselves, and every where causes a Gravitation, not according to the Surface, but according to the entire solid Content, or Quantity of the real Matter it self, contain'd in every one of those Bodies. This Property is highly remarkable, and of great Consequence against those who are unwilling to allow any real Being but Matter in the Universe. Whereas it has formerly

merly appear'd, that both the Sensitive Souls of Brutes, and the Rational Souls of Men, are *Immaterial*; and it does now most evidently appear, that the Supreme Being himself is also *Immaterial*; that is, it appears that the entire noblest Parts of the Universe, all the Living, Active, Thinking, Enjoying Parts of it, for whose Sake alone the rest of it could be made, together with the Almighty Maker himself, are really *Immaterial*. So little do the Wishes, and Desires, and Fancies, and Hypotheses of the Old and New Atheists, agree with the true System of the Universe! And so exactly do the wisest Notions, and Inclinations, and Doctrines, and Assertions of the Patrons of God and Religion, in all Ages, appear to correspond to the same! I heartily wish, that such Persons would well consider of this Observation; and no longer make use of those small Arguments, to take away the Being and Attributes of God; which still appear, when we come to real Fact, Experiment, and Demonstration, to be as little agreeable to Philosophy and Mathematicks, as they are in themselves Impious, and Blasphemous against their great Creator and Benefactor. Nature as well as Scripture teaching us, that *God is a Spirit*; and that by Consequence, *they that worship Him must worship Him in Spirit and in Truth*.

John iv 23, 24.

(14.) We learn moreover from this true System of the World, that the Supreme God, the Maker and Governor of the Universe, is a *Good* and *Beneficent* Being, and one that takes Care of the Welfare and Happiness of his Creatures. This is a direct Consequence from no small Number of the Observations foregoing, which

which shew the great Care he has taken of the Situation and Motion, and Circumstances of the several Planets, at due Distances from the Sun, and from one another, that so they might be the most commodious and comfortable Habitations for the Creatures that were to live upon them. And the same might be more distinctly and particularly observ'd as to our Earth, and its Inhabitants, if this were a proper Place for it. But this has been so frequently and fully done already, that I shall rather chuse to insert what has already been excellently said by others in the Eighth Part; or to refer my Reader to them here, than to repeat those Observations in this Place. Upon the whole, the Provision that is made by the Divine Providence for Men, and all Creatures living, which alone are capable of the Goodness and Beneficence of the Creator, both in their Entry into, and during their entire Continuance in this World, is so ample, so abundant, so surprizing, that he who carefully considers the Particulars in all Sorts and Species of Animals with us, and especially in Man, the Lord of them all, will be soon obliged not only to grant, that God is a Beneficent Being, but also, with the *Psalmist*, to Admire and Adore, and *Praise the Lord for his Goodness, and for his wonderful Works to the Children of Men.* And this Inference is the more just, because the Infinite Perfection, and Self-sufficiency of the Divine Nature, does so entirely privilege him from all Want, and from all expectation of Increase of Happiness from any of his Creatures, that every one of the Instances of his Beneficence, both in the Creation and Conservation, and Provision for those Creatures, must needs be

See Dr. Moore *against* Atheism *in Mr.* Ray's *Wisdom of God, &c.* Part II. C. 1, 2. *Dr.* Cheyne's *Philosophical Principles of Natural Religion.* Part I. C. 5. §. 24, *&c.*

Psal. 107.

be the alone free and generous Effects of his Bounty, and the gratuitous Communication of his Goodness to them; in order not to *his*, but to *their own* Happiness and Felicity; which Circumstance renders this Goodness and Beneficence the most highly Meritorious and Divine of all others whatsoever. I do not here take notice of the Objections against this Goodness and Beneficence of the Divine Nature, from the seeming Irregularities now appearing in the Moral and Rational Part of the Creation; because Divine Revelation always owns such seeming Irregularities at present, and refers the full So-

Rom. ii. 5. lution of them to the *Day of the Revelation of the righteous Judgment of God* hereafter; because we are hitherto not sufficiently acquainted with the last Result and Upshot of Things to determine those Matters; and because the most ancient *Recognitions* of *Clement* have prevented me in good Part, and afforded more Authentick and valuable Hints, than any others which we at this Day can propose, towards their Solution. To which therefore I shall here, as I did before, refer the Inquisitive Reader for better Satisfaction. Only so far the present Proposition, as well as a foregoing One, may be of use to us in that Matter; as it assures us of the Goodness and Beneficence of God, in the wonderful Contrivance of the World about us, for the Ease and Comfort of the several Creatures which are therein: Which cannot but dispose us to believe, that the same Attributes will at last appear equally glorious as to the *Moral*, as they do already as to the *Natural* World.

(15.) We learn farther from the true System of the World, that the Supreme God is invariably the

the same in all Ages; or that he is an *Immutable Being*. I do not mean that he is so rigidly *Immutable* as to be *Inexorable* to the Prayers of his Suppliants, who ask in a due manner what is fit for them; or that he has *immutably decreed* the *Fates* of Men, let them do whatever they will. These Properties would not be those of a Wise or Good Being, but of a Foolish and Evil one. Nor do the Phænomena, whence I draw this Attribute of Immutability, in the least imply any such Things. But I mean, that God appears to act by certain and constant Laws of Motion, of Gravity, of Refraction, &c. and this at all Times, and in all Places of the Universe, from the beginning of this System or World till this very Day; and from the utmost Limits of the same System one way, to those that are opposite, so far as we are ever able to examine; without any Variation from the same, from one Generation to another. Those only Cases are to be excepted, wherein God is pleased to interpose in a more immediate manner; and by leaving or contradicting the settled Course of Nature and ordinary Providence, does more effectually demonstrate his Divine Power and particular Providence in some extraordinary and miraculous Cases, for the greater Benefit of any of his Creatures. Tho' by the way it will deserve to be considered, whether even in those Cases, such miraculous Operations may not be, usually at least, brought about, rather by the means of Angels, or of some other Spiritual and Invisible Beings, which Revelation and History assures us are as his Ministers in the World, without the direct Alteration of those Fixed and Constant Laws of Nature, which we otherwise

wise find to be immutably observ'd by him. Thus that *Iron*, which was made to *swim* in Water by the Prophet, contrary to the known Law of Specifick Gravity, might as well be supported by an invisible Agent, notwithstanding its superior Gravity, as that superior Gravity, by the direct Substraction of the Divine Influence be taken away from it; and so in many other Cases; and this without the Impeachment of the Miracles themselves, or their Intentions among Mankind. However, since we have already proved that the Supreme Being is a Powerful, a Wise, a Good, and a Free Agent, as well as we are now shewing him to be an Immutable One, it cannot be just so to interpret one of these Divine Perfections, as at all to clash with, or contradict the other: and by Consequence, we are not so to establish *Immutability*, as in any Case to hinder the *free Exercise* of his Power, Wisdom, and Goodness, where sufficient Reasons determine Him to act as he pleases in the Government of the World. For that would be a strange Notion of a Powerful, Wise, Good, and Free Being, that He could exercise that Power, Wisdom, and Goodness, in only one invariable manner, let the Occasions for a different Exercise of the same be never so great or necessary. But, this Case excepted, Nature and Scripture agree, that *in God is no variableness, neither shadow of turning.*

2 Kings vi. 5, 6.

Jam. i. 17.

(16.) We learn also from this true System of the World, that the Supreme God, the Maker and Governor of all Things, is not a blind Fate, or Series of necessary Causes and Effects, but is a *Spiritual*, a *Living*, and an *Active* Being; perpetually exerting his Divine Perfections in the whole

whole Universe. I do not here take the Word *Spiritual* meerly for Incorporeal or Immaterial; which Property I have already discoursed upon; but as importing also that Vigor, and those Actions which brute Matter is uncapable of, and which are the proper Effects of Life, and on which Account the Sacred Writings so often call GOD the *Living God*, in opposition to those Dead or Inanimate Heathen Idols, which were not able to do any Thing either for Mercy or Judgment. Now that the Supreme God, the Author of all Things, must be thus a *Spiritual, Living* and *Active* Being, continually Observing, and Knowing, and Ordering, and Providing, and Acting in the World, is not only highly probable because he is the Creator of all; and therefore will not certainly forget, or overlook, or neglect the *Works of his own Hands*; but is directly proved from many of the foregoing Propositions, which have given us abundant Instances of such his Life and Activity. And no wonder that the Supreme Spiritual Being is thus Living and Active, since, as Material Substance or Body is in its own Nature, according to all the Experience we ever have of it, Dead, Inert, and wholly Unactive; so does it seem that Spiritual Substance, or Soul, is in its own Nature in general, Living, Active, and Vigorous; and that the different Degrees of such Life, Activity, and Vigor, do constitute or proceed from the several Kinds of Spiritual Beings, from the meanest sensitive Soul of a Brute, up to the Supreme Spirit of the Universe himself. I do not mean with *De Cartes*, that a Soul or Spiritual Being does for ever actually think, or is always by necessity of Nature, in Action; no more

Psal. cxxxviii. 8.

more than I believe that a Body is ever really, and by necessity of Nature in Motion; but that it is always Quick, and Active, and Ready, upon all Occasions and Objects that present themselves, to think, and consider, and resolve, and exert it self; and that this Faculty is one of its main Distinctions from Body or Matter, which is entirely destitute of all such Faculties whatsoever. And certainly, He that produc'd all the Living, and Active Powers and Abilities which are in the Creatures, must himself possess them in the highest and most exalted manner possible; which Life and Activity, if He, the Supreme Being, were wholly bereaved of, he would be so far from the Object of our Worship, Fear, Love, and Adoration, as all Men naturally look upon him to be, that he would be certainly the Subject only of Neglect and Contempt. Wherefore Philosophy, as well as Religion, assure us of the Truth of what *Darius* an Heathen King was once obliged to acknowledge, that *Men ought to tremble and fear before God; for He is the Living God, and stedfast for ever; and his Kingdom that which shall not be destroyed. And He worketh Signs and Wonders in Heaven and in Earth.*

Dan. vi. 26, 27.

(17.) We farther learn from the true System of the Universe, That the Supreme God, the Maker and Governor of all Things, is but *One*. This *Unity* of God, or that there is but *One* Infinite, Eternal, Omnipotent Being or Agent, who created and governs all Things, is among the first Dictates of Nature and right Reason, when it reflects upon the obvious Phænomena of the World, and is most clearly confirm'd by all the foregoing Observations. The *Universe*

appears

appears thereby to be evidently *One Universe*; govern'd by *One Law of Gravity* through the whole; and observing the *same Laws* of Motion every where. The *Conduct* of the whole is every where *One and the same*; and not the least Signs or Traces do appear of any Opposite, or Coordinate Power interposing it self in any of its regular Phænomena. So that this *Unity of God*, is now for ever established by that more certain Knowledge we have of the Universe; as it was naturally also, tho' in a lesser Degree, discoverable by all Mankind before. So true and just are the Words of the Apostles in their *Constitutions*, concerning the Occasion of God's imposing on the *Jews* the rigid Ceremonies of the Law of *Moses*, after their Idolatry about the Golden Calf " I impose these Things on thee, - - - that be- " ing press'd and gall'd by thy Collar, thou mayest " depart from the Error of Polytheism, and " laying aside that, *These are thy Gods O Israel*, " mayest be mindful of that, *Hear O Israel, the* " *Lord our God is One Lord*; and mayest run " back again to that Law, which is inserted " by me in the Nature of all Men, *That there* " *is only One God in Heaven, and on Earth*.

L. VI. C. 20.

Exod. xxxii. 4. Deut. vi. 4.

(18.) Lastly, we learn from this true System of the World, that we and all Creatures are entirely under the *Dominion*, and subject to the supreme *Authority* of this One God, as to *our Lord and Governor*. This is the most natural and obvious Consequence, from what has been hitherto proved, concerning the Existence, the Attributes, and the Providence of the One Supreme God; that this World, wherein we all live, is *God's World*; that this System of the Universe,

Universe, is God's great House, or Family, or Kingdom; and that all Rational Beings are God's Creatures, the Members of that his Family, and Subjects of that his Kingdom; owing all possible Obedience, Duty, and Homage to him, as to their great Master and King. This absolute *Supremacy* of the One God, and Obligation of all Derived and Created Beings to entire *Submission* to him, is so plain from these Considerations, that as He must be the most *egregious Fool*, in point of Prudence, that will dare to oppose himself to the Omnipotence of his Almighty Lord and King; so is he the most *profligate Wretch* in point of Duty, that will venture to refuse his Submission to the Infinite Wisdom and Goodness of his most beneficent Creator, Father, and Benefactor, *In whom he lives, moves, and has his Being; and whose Offspring he is.* Nor will Philosophy it self in the least scruple the justness of that Exhortation made use of by the *Psalmist* upon this Occasion; *Because the Lord is a great God, and a great King above all Gods; because in his Hand are the deep Places of the Earth, and the Strength of the Hills is his also; because the Sea is His, and He made it, and his Hands formed the Dry Land; O come therefore, let us Worship and Bow down, let us Kneel before the Lord our Maker: For he is our God, and we are the People of his Pasture, and the Sheep of his Hand.*

Acts xvii. 28.

Psal. xcv. 3-7.

PART

Part VII.

Important Principles *of* Divine Revelation, *confirm'd from the foregoing* Principles, *and* Conjectures.

SINCE it has now pleased God, as we have seen, to discover many noble and important Truths to us, by the Light of Nature, and the System of the World; as also, he has long discovered many more noble and important Truths by Revelation, in the Sacred Books; It cannot be now improper, to compare these two *Divine Volumes*, as I may well call them, together; in such Cases, I mean, of Revelation, as relate to the Natural World, and wherein we may be assisted the better to judge, by the Knowledge of the System of the Universe about us. For if those Things contained in Scripture be true, and really deriv'd from the Author of Nature, we shall find them, in proper Cases, confirm'd by the System of the World; and the Frame of Nature will, in some Degree, bear Witness to the Revelation. But if those

Points contain'd in Scripture be false, and did not really come from God, we shall find the Frame of Nature, which is now much better understood than in the Days of those Antient Writers, frequently to contradict them, and so to detect their Forgery, and prevent their further Imposition upon Mankind. Nor indeed have we any way so sure and satisfactory to examine the pretended very ancient Accounts of Things by, as this in general; I mean, the trying their Verity in such Points, whether Natural or Historical, as we have sure Methods of knowing, whether the Things asserted in them were, or might be really true or not. For as a plain Disagreement of Nature, or certain History, from Scripture, in such Points, will afford a terrible Suspicion, that the latter is either false, or at least interpolated; so will as plain an Agreement be a mighty Evidence for the Truth, and Uncorruptness of those Scriptures; and this even in general, as to such other Contents of the same, as can no way come under the like Methods of Examination. If I am once fully satisfy'd, that a Witness is Upright and Honest, even in several Points where there was the greatest Suspicion as to his Sincerity, he will deserve the better Credit in other Cases, even where no corroborating Evidence can be alledg'd for his Justification. To this kind of Evidence then do I Appeal on behalf of those Sacred Writings; and do plead for their Reception, as Genuine and Authentick Records, in the several particular Cases following.

(1.) The Scriptures agree to declare the very same, *all the same Truths*, and ascribe the very same, *all the same Attributes* to God, which

we

we have shewed to be real, and to belong to him from the Consideration of the System of the World.

(2.) Their Accounts of the *Creation* of the World out of a *Chaos*, when rightly understood, are not only agreeable to many Remains of Ancient Prophane Tradition, but to the Frame and Laws of Nature, and the true System of the Universe.

(3.) Their Chronology, or Accounts of the *Antiquity of the World* are right, and agreeable to the best Methods which Nature and Philosophy afford us for the discovery of the same.

(4.) Their Accounts of the *Paradisiacal State*, and its Mutation from that State, before the *Antediluvian One*, after the Fall of Man, are agreeable to Nature, and the true Frame of the World.

(5.) So are their Accounts of the *Universal Deluge* in the Days of *Noah*.

(6.) So are their Accounts of the *Invisible World*, or of *Angels* and *Dæmons*, their Places and Ministrations.

(7.) So are their Accounts of the future *Conflagration* of the World.

(8.) So are their Accounts of the *Resurrection of the Body*, and *Renovation* of Things.

(9.) So are their Accounts of the future *Consummation of all Things*.

(10.) So are their Accounts of Ἅδης, or of the Place of Departed Souls, in the intermediate State, before the Resurrection.

(11.) So are their Accounts of *Heaven*, or of the Place and State of Happiness for Good Men after the Resurrection.

(12.) So are also their Accounts of *Hell*, or of the Place and State of Punishment for wicked Men, after the Resurrection.

These are the Heads I propose to Discourse of. Only before I begin, I must, as I have already done under the Head of Conjectures, to which this Part will commonly bear a very near Relation, intreat the Reader to distinguish exactly between the foregoing *Propositions*, with their *Inferences*, as to Natural Religion, Chapters IV. and VI. which I propose as *certain*; and the foregoing *Conjectures*, with their *Inferences*, as to Reveal'd Religion, Chapters V. and VII. which I propose as *Probable* only. I do not mean, that the Doctrines themselves, contain'd in Scripture, are only Probable; for I believe they have the Certainty of Divine Revelation it self: But that the Accounts, or Natural Solutions of those Points here offered, which are commonly the Result of my own peculiar Thoughts and Notions, upon the Comparison of Scripture and Philosophy together, are not yet to be look'd on as more than private or *probable Conjectures*, humbly proposed to the Consideration of the Publick. I speak this in the general, and with regard to the greatest number of them: For as for some few of them, especially that relating to the Deluge of *Noah*, I cannot but look on them as so exceeding probable, that I can scarcely avoid placing them under the foregoing Affirmations, or Assertions. But to come to Particulars,

(1.) I observe that the Sacred Accounts declare the very same, *all the same Truths*, ascribe the very same, *all the same Attributes* to God, which we have shewed to be real, and to belong

long to God, from the Confideration of the true Syftem of the World. This I have taken care to prove or illuftrate all the way as I went over thofe particular Truths and Attributes; and took notice that both Nature and Scripture all along correfpond to the feveral Conclufions: That they jointly agree that the Souls of Brute Creatures are diftinct from their grofs Bodies, and are fenfitive, and moft probably *incorporeal*; That the Rational Souls of Men are more certainly and entirely *Immaterial*; that they are alfo *Immortal*, or will naturally exift after the grofs Bodies are diffolved; that there is a God or *Supreme Being*; that this World has not been *Eternal*, but was *Created* by that Supreme Being; that this Supreme Being, the Creator of all Things, is an *Eternal* Being; that He Exercifes a continual *Providence* over the Creatures he has made; that he is not a *Neceffary*, but a *Free Agent*; that He is *Intelligent* and *Omnifcient*; that He is an *All-wife*, and that He is an *All-powerful* Being; that He is *Omniprefent, Immaterial*, and *Immutable*; that He is *Good* or *Beneficent*; that He is a *Spiritual*, and *Living*, and *Active* Being; that He is but *One*; and that He is therefore the alone Supreme *Lord* and *Governor* of the whole Univerfe. Now this cannot but be a great Confirmation of the Sacred Records, that fuch Deductions from Aftronomy and Natural Philofophy, as their Authors could either not at all, or very imperfectly make when they lived, do yet, upon that vaft Improvement of Natural Knowledge, which has been of late made, all appear right and agreeable to the true Syftem of Things. And this is more particularly to be obferved, as to fuch of the be-
fore-

fore-mentioned Divine Attributes, as the Heathens differ'd from the *Jews* and Christians in; wherein the System of the Universe does, as we have seen, every where bear Witness against the Heathen, and for the *Jewish* and Christian Notions in those important Matters. I instance distinctly in that grand Doctrine of the *Unity* of God, wherein these Sacred Records did all along differ from the current Notions, and consequent Practices of the rest of the World, in all those Ages whereto they belong; and indeed insisted on it to that degree, as to lay their entire Foundations on that *Unity of God*, in the directest opposition to the *Polytheism* of the rest of the World; and this without the Pretence of greater Natural Knowledge than was in other Nations, to derive that Notion from; and yet with such Boldness and Assurance, as to Ordain the hazard of Life it self, and of all the Comforts thereof upon its Truth, and the Hopes derived from it. This Agreement of the System of the World to the *Unity of God*, now so certainly discovered by the System of the *Universe*, in exact Concord with the *Jewish* and Christian Records; and in direct Contradiction to the Opinions of the generality of the Heathen World, even in the Politest and most Learned Ages of it, is like a *Solemn Determination* upon an *Open Appeal* made on the sides of the *Jews* and Christians, against all the rest of the World, in this most Important and Fundamental Point of true Religion. Which Determination ought therefore to be own'd as the highest Attestation to that true Religion, and those Original Records wherein it is contain'd, that could easily be desir'd or expected:

expected: And which accordingly ought to be allowed to be of the greatest Weight and Moment possible in the present Case. I observe,

(2.) That the Sacred Accounts of the *Creation* of the World out of a *Chaos*, when rightly understood, is exactly agreeable to the Frame of Nature, and the true System of the Universe. That Account of the *Mosaick* Creation which I mean, is this, That, taking the only Example which is in Nature of a real *Chaos*, I mean the *Atmosphere of a Comet*, for the *Mosaick Chaos*; and taking that Chaos or *Atmosphere*, not to be revolv'd about its own *Axis*, till after its Formation into a Planet, which Nature fully agrees to, and which would imply a *Day* and a *Year*, to be all one, that " This *Mosaick* Cre-
" ation, I say, is not a Nice and Philosophical
" Account of the Origin of all Things; but an
" Historical and True Representation of the
" Formation of our single Earth out of that
" confused Chaos, and of the successive and
" visible Changes thereof each Day, or Year,
" till it became fit for the Habitation of Man-
" kind. Now as to this Matter, I have entirely prevented my self in my Preface to the *New Theory of the Earth*; wherein I have at large Discoursed thereof; and I hope, in good Measure, to the Satisfaction of the Learned and Inquisitive; whither therefore I must in this Place refer my Readers. Only I shall desire them to reflect here upon a few Things in Natural Knowledge thereto relating, which are peculiarly suitable to my present Purpose; as highly confirming the Truth of the Sacred, or *Mosaick* History of the Creation. The Things
I mean

I mean are these Four; First, Nature does now exhibit to us such a *Chaos* as that History supposes, for the Fund and Promptuary of our Earth; which till lately the World could not know, but by Revelation: Secondly, Nature does now shew the Possibility of the greater Length of the Periods, or Days of Creation, than those of our ordinary Days, without which no Natural and Rational Account could be given of that Creation; and this consistently with the Letter of *Moses*, that those Periods were still, strictly speaking, *Days* at the same Time: Thirdly, Nature does now so clearly shew the Impossibility of the *Eternity* of this System, and much more of this particular Earth, in its present State, as prepares the way for the Belief of that Sacred Account of its Original Creation: And Fourthly, Nature does so plainly shew the necessary Interest the Supreme Being has in all the ordinary Appearances of Nature, and in the Preservation of the several Worlds now in being, as much more implies his Interest in the first Original, and Primary Settlements of the same; and such as prepares us most easily to believe what *Moses* asserts of the Interposition of *Divine Power*, and *Wisdom*, and *Goodness*, in that whole Affair. These Observations deserve a peculiar Regard; and Astronomy does now, in these Respects, fully support and attest to the Sacred Accounts, thus reasonably understood, against all the opposite Notions of those ignorant or prejudic'd Philosophers, who pretended to give different Accounts of such Matters; and especially against those Two Famous Antiscriptural Hypotheses of the *Eternity* of Things; and of their Original from meer *Chance* and *Accident*.

Accident; both which, as we have already shewn, are entirely confuted, and rendred not only incredible, but ridiculous, from the present Knowledge we have of the true System of the Universe. I observe,

(3.) That the Sacred *Chronology*, or Scripture Account of the Antiquity of the World, and its Duration since the Deluge, is right, and agreeable, not only to the most certain Remains of Ancient Prophane History, but to the best Methods which Nature and Philosophy afford us for the Discovery of the same. I mean not only, that it appears from the true System of the World, that neither the Whole, nor any Parts of the same, can have been strictly *ab æterno*, in their present Condition, of which before; but that the Antiquity of our present Earth, and it's Inhabitants, whether taken from Ancient Prophane Tradition, and History, or from the Phænomena of Nature, does best agree to the Sacred Accounts, which make it nearly 6000, or 7000 Years Old, and not more. Now as to this Matter, I must own that a Person of very great Sagacity, has lately advanc'd a contrary Notion, from the Degree of Saltness of our present Ocean, compar'd with the Length of Time which he supposes necessary, according to his own Hypothesis, to bring it to that Degree. Whence he seems to think it to be much longer, not only than the Scripture Account of the Time since the Deluge; beyond which, I should think it in vain to extend any such Calculations; but much longer than the Time since the *Mosaick* Creation it self, as delivered in the same Scripture; unless we extend the Duration of the Days of Creation to an immense Length, without

Philos. Transact. N°. 345.

out any Authority for so doing. But then, this Person allows all this to be built on an Hypothesis, an uncertain Hypothesis, of the Origin of the Saltness of the Ocean; and he allows that such Trials have not been made, nor indeed are now possible to be made, as are necessary even on this Hypothesis, to determine that Duration, in this or several future Ages. So that at the bottom the whole is, by his own Concessions, very uncertain, and only built on some *Suspicions*, which it cannot be now known whether they have any Foundation in Nature or not. In which Case, by the way, it had been but reasonable to avoid the giving any Intimations to weaken the Sacred Chronology, unless there had been some real or sure Evidence against it; which yet is not here pretended to. However, leaving this Notion, as not yet worthy of a direct Confutation, I venture to assert, that the best ways of Determination we now have of the Age of the World, whether from Prophane History, or the Phænomena of Nature, do very well agree to the Sacred Chronology, and confirm us in the Belief that the Earth has not been in its present State above 6000 or 7000 Years; and that since the general Flood, there have scarce yet passed 5000 Years; according to the Sacred Chronology thereto relating. Now the best ways of knowing this Duration of the World, abstractedly from Scripture, seem to be these Five following. (1.) By Ancient Prophane Histories, directly relating to such Matters. (2.) By the Histories of the Beginning and Progress of Arts and Sciences. (3.) By the Accounts of the Origin and spreading of the several Colonies of Nations

Nations all over the World. (4.) By the present Number of Mankind upon the Face of the Earth, compar'd with the best Computations we have of the Time necessary for such their Increase and Doubling. (5.) By the present State of the Celestial Motions, and Terrestrial Appearances, and the Length of Time necessary for any such Irregularities arising therein, as would be sensible to us. As to the former Four Methods, which justly pretend to the greatest Nicety, they are not proper for this Place; and they have been so fully examin'd already by others, or by my self elsewhere, and found so well to agree to the forementioned Intervals, that I believe judicious Men must pronounce upon the whole, they are considerable Attestations to this Sacred Chronology. But as to the last Method, the Consideration of the Celestial Motions, and Terrestrial Appearances, I do not know that they can determine to any Degree of Exactness, how long they have continued in their present State. Only thus far, that none of those Motions or Phænomena do contradict the Sacred Accounts, and that they all shew that the World cannot have been very much older than those Accounts affirm. Thus the Moon's Menstrual Motion must be gradually stopped by the Æther, or fine Medium in which it revolves about the Earth: But then this is hitherto so perfectly Insensible, that Dr. *Halley* seems to have been the first that discovered any Occasion in Astronomy, for making the least Allowance for any Inequality in that Motion. Thus the Earth's and Moon's Annual Motion must be gradually retarded by the same Resistance; yet so small has this hitherto been, that

Bochart Phaleg. L. C. J. Hale's *Origination of Mankind.* See Chronol. Old Test. *p.* 60, -68.

Sir Isaac Newton's *Princip.* 2d *Edit. p.* 481.

that the Astronomers have not yet observed it. Thus the Consequence of the Retardation of the Earth's Diurnal Motion; which must also in length of Time arise from the same Resistance, would be the receding of the Sea from the Dry Land in the Torrid, and its overflowing it in the Frigid Zones: [unless the Earth be fluid within, and so accommodates it self to such Alterations] which is not yet in the least observable. Thus the Fluids of our Earth are found gradually to diminish; yet is this so very inconsiderable hitherto, as no way to be distinctly found by any Inconveniences arising from it. Thus the Mountains do, for certain, wash away, and are diminish'd; and the Valleys receive what the others lose, and are augmented: Yet do not we hitherto observe any notable Inequalities arising therefrom. In short, all such defects, decays, or irregularities, which must in length of Time, according to the settled Laws of Nature, arise in our present Constitution, appear to have been hitherto so very small and inconsiderable, since the beginning of this Settlement, that we have thereby certain Evidence that its Age cannot be very much greater; and no Evidence that it is at all greater than what is contain'd in the Sacred Chronology. Which Thing, how considerable a Confirmation it is of that Chronology, I leave to the Impartial to determine. I observe,

(4.) That the Sacred Accounts of the Mutation from the *Paradisiacal State* before, to the *Antediluvian one*, after the Fall of Man, is agreeable to the State of Nature, and the true Frame of the World. That God, when he first made Man upon this Earth, placed him in Paradise;

Paradife; that He there gave him a Law for the Tryal of his Obedience; that Man did tranfgrefs that Law; that this Tranfgreffion was occafion'd by the Temptation of the Old Serpent the Devil and Satan; and by the firft Compliance of the Wife, and by her Perfuafion of her Husband; that they were thereupon caft out of Paradife, and the happy State of Nature was much altered for the worfe; that the Serpent was accurfed, and fubjected at laft to the Power of the Seed of the Woman; that the Ground was alfo curfed, and made to bring forth Thorns and Bryars, and not to bring forth its Fruit without the extraordinary Toil and Sweat of Mankind; that the Female fhould be in greater Subjection to the Male than otherwife fhe fhould have been, and than many other Females now are; that fhe fhould have greater Sorrow in the Conception and Bearing of her Offspring than otherwife fhe fhould have had, and than any other Females now have; and that ever afterwards the Race of Men fhould have alfo fuch a Senfe of Shame, or *pudor circa res venereas*, as they had not before, and as is not in other Animals; and withal more conftant Inclinations that way than thofe Animals have; all which has come to pafs accordingly, as ftanding Memorials of the Truth of this Ancient Hiftory of the Original State and Circumftances of Mankind. Accordingly, for this we have not only thefe natural Tokens, but the certain Affirmation of the *Mofaick* Hiftory, as all along fuppofed and confirmed in the Heathen Traditions, and in the *Jewifh* and Chriftian Revelations, as has been fhewn by others. But then, what the Alteration in the State of the Earth and of Nature could

Gen. i, ii, iii.
See the Commentators there.

could be, the Philosophy and Astronomy of former Ages was utterly at a Loss to determine, till upon the Consideration of that Matter, and the Comparison of Scripture and Nature, now better understood, together, I discovered that the commencing of the Earth's Diurnal Rotation, not at the beginning of the *Mosaick* Creation, but after the Earth's compleat Formation, and after the Fall of Man, would give the Best, the most Rational, and Philosophical Account of this Alteration of all other; and I discovered, that this later Time of its Commencement, would best agree, not only with the Sacred Accounts, but with the Ancient Profane Traditions, and with the present Phænomena of Nature also. All which I have long ago discoursed of in my *New Theory of the Earth*; to which I must therefore refer my Reader for satisfaction. Only I shall desire him in particular to take notice of what is there especially insisted on under this Head, and which is peculiarly proper to my present Purpose, and that is, Dr. *Halley*'s Inferences from the Variation of the Needle's Declination from the *North*, and of Mine from the *Mosaick* History of the Creation and Fall of Man, relating to the Time and Circumstances of this Commencement of the Earth's Diurnal Revolution; and thence to observe how exactly they both agree together. Nor shall I enlarge farther here upon that whole Matter, but leave it entirely to the Determination of the Judicious. I observe;

Hypoth. iii. p. 85-- 118.

2d *Edit*. p. 109, 110.

(5.) That the Sacred Accounts of the *Universal Deluge*, in the Days of *Noah*, is exactly agreeable to the State of Nature, and to the true Frame of the Universe also. Indeed the Solution of this moft remarkable Phænomenon of an *Univerfal Deluge*, with its moft numerous and eminent Circumftances, as defcribed in the *Mofaick* Hiftory, which till this Age could no way be folved in a Natural way, nay feem'd utterly uncapable of any Philofophical Solution at all; is now, I think, become fo plain, evident, and certain, from the Phænomena of Comets, with their Atmofpheres and Tails, now fully difcovered; efpecially from the particular Circumftances, and Periods of the laft moft famous Comet of 168$\frac{0}{2}$, which appears to have been the Phyfical Caufe of the fame Deluge: I fay, the Solution of thefe wonderful Phænomena's, as given in the Second Edition of my *New Theory of the Earth*, with its additional Sheet, is become now fo plain, evident, and certain, that I own I cannot but be my felf very much furprized and fatisfy'd with it. and equally furprized and fatisfy'd with that ftrong Confirmation it affords to the Sacred Records, in one of the leaft probable, or moft exceptionable Branches thereof. This is too large a Subject to be duly treated of in this Place. But I beg of the Inquifitive Reader, that he will carefully weigh the very probable Solution I long fince gave of that Deluge of *Noah*, from the Approach of a Comet, before I fo much as hoped for the Knowledge of. that very Comet which did fo approach to it, and did caufe the fame; and that he will confider the ftrong additional Evidence fince arifen from the knowledge of that very Comet which did certainly approach to us on or about that very Year, and on or about that very Day of the Year, when the Sacred and Profane Accounts agree that Deluge began. If this be not fatisfactory

See that Sheet here at the End.

factory Evidence in such a Case, I do not well know what Evidence will be so esteem'd. For my self, I must profess, that while I look on the Solution of the other Phenomena under this Chapter as remarkable enough, and commonly not a little probable also; I cannot but look on the Solution of the Deluge by that very Comet, which I my self saw *A. D.* 1680, 1681, to be in a manner certain; and by Consequence I cannot but esteem the Evidence thence arising, for the Truth of the Sacred History in this important Case exceeding strong and satisfactory. Nor do I think, that so unexpected and eminent an Attestation, as that of the Circumstances and Period of this Comet, for solving the Deluge, lately discovered, most certainly is, has ever happend to any so strange an Hypothesis before, since the World began; which Thing cannot but be highly pleasing to my self; and I think is highly worthy of the Observation of others also. I observe,

(6.) That the Sacred Accounts of the *Invisible World*, or of good Angels, and wicked Dæmons, their Places and Ministrations, is exactly agreeable to the true System and Phænomena of Nature. Now that the Sacred and Prophane Accounts, and the Histories of all Nations and Ages, do suppose and inform us, that besides the Supreme Invisible Deity, besides the Visible Material World of Bodies, and besides the Invisible and Incorporeal World of Souls or Spirits, inhabiting in Visible and Gross Bodies, there are also another Species of Beings belonging to our System: I mean, those Souls, or Spiritual Beings, who are either wholly free from Bodies, or rather free from such Gross and Visible Bodies as we have, but inhabit ordinarily in purer and more etherial Regions, in more subtle and aerial Bodies or Vehicles; who are Invisible generally, and Intangible to our gross Senses, but not wholly Incorporeal, or unconcerned with us and our Affairs here below; that, I say, the Ancient, Sacred and Prophane Accounts, and Histories, do assure us of the Existence of these Species of Beings, every body that has ever read either Ancient or Modern Books, cannot deny or doubt. But that present Nature does at all favour these Accounts, and that we can give the least Evidence from the Phænomena of the World, of their Being, or Place of Abode, or Influence here below, is what is not generally pretended to by even the Christian Philosophers. Now in this Case I shall venture a Step farther, and shall here set down such Observations from Nature and Astronomy, as seem to me to favour this Account of the Invisible World. Accordingly I observe, (1.) That Nature does as fully allow of the Existence of Spiritual and Invisible Beings *out* of gross Bodies, as *in* them. Nor can those who are convinc'd by the Phænomena of Animals, both Irrational and Rational, that they are compounded of Immaterial and Invisible Souls, as well as of Material and Visible Bodies, as we have already discours'd, at all scruple the Existence of such Spiritual and Invisible Souls, either by themselves, or united to much finer Bodies than those of our gross Animals here below. I observe, (2.) That Nature does favour the Existence of such Creatures, by shewing us such large and noble Regions of the World,

World, as best of all suit the Habitation of such Beings; and which, if there be not such Beings, seem, contrary to the usual Case of the other Parts of the System, to be wholly destitute of Inhabitants: I mean all the wide Spaces of the Atmospheres of the Planets, especially those still, calm, clear, and æthereal Regions of the same which are above the Clouds, and Storms, and Disorders of their lowest Parts. Nature, as we still find, abounds in all proper Places, with Living Creatures, not only on the Earth, or dry Land, but within the Earth, and Waters, and lowest Air, every where; all made to enjoy their Creator's Bounty, and to be serviceable to other Beings Superior to themselves. What Reason can there then be to suppose that this Air, the noblest Fluid in the Universe, even in its purest and most Celestial Parts, should be destitute of Living Inhabitants? which yet it must be, in case we exclude those *Invisible Powers* above-mentioned. Now, in order to shew how little Prejudice to the Existence of such Beings, that Circumstance of their being *Invisible*, ought here to be; I observe, (3.) That whatever proper Inhabitants the Air has, their very Bodies ought to be *Invisible*, because the Air it self, whereof we may suppose them made, is ever so. This is the wonderful Property of Air, strictly speaking, and that Property, which among all Corporeal Beings, otherwise sensible, seems peculiar to it, that it is ever, upon the utmost Condensation possible, absolutely to us *Invisible*. Whence 'tis no wonder, that all such Beings as live in it, and perhaps take their Bodies from it, how real or powerful so ever, are
like-

likewise ordinarily *Invisible* to us. I observe, (4.) That the known Phænomena of our Air, seem in a peculiar Manner, to require and suppose the Existence and Agency of such Invisible Beings therein, as we are here speaking of; and cannot be fairly and mechanically solv'd without them. This I have particularly taken Notice of, in my Account of the unusual Meteors lately seen in our Air, to which I refer the Reader. Nor do I find that any other Philosophers are able to give us a better Account of those Aerial Phænomena, without that Hypothesis. So I cannot but conclude, that the Appearances of Nature do in this, as well as in other Cases, attest to what Discoveries Divine Revelation has made relating thereto: And that there are Good and Bad Spirits in the Air, in our Neighbourhood, ready to perform what the Sacred Writings ascribe to them. As to the other Place allotted for certain to some, now to us, Invisible Beings, in Scripture, I mean *in the Heart of the Earth*; I have already made it probable from Natural Philosophy, that the Heavenly Bodies have such mighty Cavities within, as are the proper Receptacles for some such Beings; and shall not need here to enlarge on that Subject; especially since it will come again to be considered under the Tenth Particular hereafter. I observe,

See Dr. Halley's Account of this Phænomenon, Transact. Philosoph. Nº. 347.

(7.) That the Sacred Accounts of the Future *Conflagration* of the World, is exactly agreeable to the true System of the Universe. Now this Conflagration of the upper Earth, and all that is upon it, according to the Discoveries made by Divine Revelation, is so natural, or rather

2 Th. f.i.8.
2 Pet. ii.
2, 10.

necessary a Consequence of the Approach of a Comet to a Planet, when it has just been broiled in the Sun's Rays, which Astronomy now knows to be not only possible, but, in the course of Time, sometimes certain; and I have so fully prevented my self upon this Head in my *New* *2d Edit. p. 440, -449.* *Theory of the Earth*, that I shall not need to enlarge upon it here: I only beg of the Inquisitive Reader to observe, That the Two grand Catostrophes which the Scriptures certainly make our Earth alone subject to, I mean a *Deluge*, and a *Conflagration*, are those Two Catastrophes, and those Two only which the true System of the World shews such Planets to be naturally liable to: Which Observation, how great a Confirmation it is of the Truth and Divine Original of those Scriptures, I leave to every Reader's own serious Consideration. I observe,

(8.) That the Sacred Accounts of the *Renovation of Things*, and of the *Resurrection of the Body*, are very agreeable to some known Phænomena of Nature. That the παλιγγενεσία, or *Renovation*, not utter *Destruction* of the World, may be the Natural and Regular Consequence of that *Conflagration* we have been lately speaking of, I have particularly shew'd towards the *Ubi suprà.* End of my *New Theory*; and shall not need to repeat it in this Place. But that the very *Resurrection of the Body* should have any thing properly to Countenance it, or resemble it in Nature, will not be easily believ'd. Yet do I *See Mr.* look upon those Changes which are made in the *Derham's* Bodies of not a few Insects, as particularly in *Physico-* the Silkworm; while still upon the most sur-*Theology.* prizing Mutations in the Body, the same Life or Soul is the Inhabitant within; as no ill Resemblances

semblances of, or preludes to, the nobler Wonder of the Resurrection of Human Bodies. I mean this, as we thence learn how very different Forms and States the same Creature can naturally put on, without being really another Creature: Just as the Christian Religion informs us the Souls of Men must undergo in the several Conditions or Periods (1.) Of this gross Earthly Body now. (2.) Of the Aereal Vehicle in the intermediate State. And (3.) Of the Spiritual Body after the Resurrection. Nor should we our selves have been able to believe such Mutations in the same Insects to be true, unless continual Experience did assure us thereof: as neither does the Christian Religion expect the Belief of the other, but upon the Attestation of Him who made all those mutable Creatures, with Man also, who is to undergo those other more important Mutations. *See Constitut. Apost. V. 7.*

(9.) I observe, that the Sacred Account of the future *Consummation of all Things*, is agreeable to the true System of the Universe. What I here mean is that particular and final Catastrophe of our Earth, spoken of by St. *John* in his *Revelation*, where he informs us, that *He saw a great white Throne, and Him that sat on it; from whose Face the Earth and the Heavens fled away, and there was found no place for them*: Or when a final Period is to be put to the present Place and Use of this Earth, with its Atmosphere. Now that this Catastrophe may naturally and regularly befal our Earth, or any of the Planets, according to the true System of the World, and without a Miracle, I have already observ'd in the last Proposition of my *New Theory*. Nor can any one that knows how a Comet *Apoc. xx. 11.*

Comet may regularly strike against a Planet in its Course, and thereby remove it from its present Station, into an Orbit of a quite different Situation, Nature, and Use, from what it now has, make any difficulty at all in believing this, *viz.* that there will be at last, a *Consummation of all Things* belonging to this Sublunary World, according as the Inspired Writer has delivered it to us. I observe,

(10.) That the Sacred Accounts of ᾍδης, or of the *Place of departed Souls*, in the intermediate State before the Resurrection, is very agreeable to the true Frame of the Universe. That ᾍδης, or the Place of departed Souls, till the Resurrection, is either in the Air, or in the Heart of the Earth, seems to me the Importance of all the Ancient and Sacred Records we have of that Matter; *i. e.* they seem to me to imply, that some of them are at liberty in the Air, and others imprison'd in the Earth; which two Places we have shewn to be, Philosophically speaking, the only fit Places for their Habitation also. So that hitherto Nature and Scripture seem to me entirely to agree, and to bear Witness mutually to each other in these Matters.

See the Appendix to *My Boyle's Lect. or my Serm. and Essays.* P. 170--178.

(11.) I observe, that the Sacred Accounts of *Heaven*, or of the Place and State of Happiness for Good Men before the Consummation of all Things, is not only agreeable to the Remains of Ancient Profane Tradition, but to the true System of the World also. This happy State is describ'd in Scripture to be a *State of Light*, a *Reward in Heaven*, and introduc'd by Good Men's *meeting the Lord in the Air*; and so being *ever with the Lord*. Which if it be after the Conflagration, as seems not improbable, will

Ps. xxxvi. 9.
Coloss. i. 12.
Luc. vi. 23.
1 Thess. iv. 17.

of RELIGION.

will belong to a more pure and purged State of our Air or Heaven than what we now enjoy; which will well agree to such *Glorious* and *Spiritual,* and *Uncorrupt* Bodies as Good Men are to have at that Time. Wherefore, as we have already seen, that the Air in its present State, both according to Nature and Scripture, is one proper Place for Invisible Beings, those I mean that inhabit Aerial Bodies, so may it by the *purifying Fire* of the Conflagration be so meliorated as to be a proper Place for the Habitation of the Pious, with their Spiritual Bodies also, till the Consummation of all Things. For as to the State and Place of Happiness after that Consummation, I do not know that the Sacred Writings afford us any particular Light about it; and so I shall not presume to indulge my self in any groundless Conjectures thereto relating; as content with this exact Agreement of Nature and Scripture till this grand Period of our World, which seems to me to be the Grand Period of those Sacred Books also. [1 Cor. xv. 42, 43.]

(12.) I observe, that the Sacred Accounts of *Hell,* or of the Place and State of Punishment for wicked Men after the general Resurrection, is agreeable not only to the Remains of ancient profane Tradition, but to the true System of the World also. This sad State is in Scripture describ'd as a State of *Darkness,* of *outward Darkness,* of *blackness of Darkness,* of *Torment* and *Punishment for Ages,* or *for Ages of Ages, by Flame,* or *by Fire,* or *by Fire and Brimstone, with Weeping and Gnashing of Teeth;* where *the Smoak of the* Ungodly's *Torment ascends up for ever and ever*; where they are Tormented *in the Presence of the Holy Angels,* [Matt. viii. 12. Jude 5. 13. Matt. xxv. Luc. xvi. 24. Matt. viiii. 12. Apoc. xiv. 10, 11.]

and

and in the *Prefence of the Lamb* ; *when the Holy Angels shall have separated the Wicked from among the Just, and have cast them into a Furnace of Fire*. Now this Description does in every Circumstance, so exactly agree with the Nature of a Comet, ascending from the Hot Regions near the Sun, and going into the Cold Regions beyond *Saturn*, with its long smoaking Tail arising up from it, through its several Ages or Periods of revolving, and this in the Sight of all the Inhabitants of our Air, and of the rest of the System ; that I cannot but think the Surface or Atmosphere of such a Comet to be that *Place of Torment* so terribly described in Scripture, into which the Devil and his Angels, with wicked Men their Companions, when delivered out of their *Prison* in the Heart of the Earth, shall be cast for their utter *Perdition* or *second Death* ; which will be indeed a terrible but a most useful Spectacle to the rest of God's rational Creatures ; and will admonish them above all Things to preserve their Innocence and Obedience ; and to *fear him who is* thus *able to destroy both Soul and Body in Hell.*

Mat. xiii. 49, 50.

2 Pet. ii. 4.
Jud. v. 6.
Apoc. xx. 14.

Mat. x. 28.
Luc. xii. 5.

Part VIII.

Such Inferences shewn to be the common Voice of Nature and Reason, from the Testimonies of the most considerable Persons in all Ages.

N. B. THE Testimonies under this Head will be of themselves so plain, that I shall need to make no Comment nor Reflections upon them; but then they will be withal so numerous, that I must beg the sober Reader's Patience and Attention in the distinct Perusal and Consideration of them: Which certainly are but a due Debt to the Importance of the Subject, and to the Character of the Witnesses. I begin with the Book of *Job*; which I esteem the Ancientest Record now extant in the World.

Job] God is wise in Heart, and mighty in ix. 4- 10. strength: Who hath hardened himself against him, and hath prospered? Which removeth the Mountains, and they know not: which overturneth them in his Anger. Which shaketh the Earth

out

out of her Place, and the Pillars thereof tremble. Which commandeth the Sun, and it riseth not: and sealeth up the Stars. Which alone spreadeth out the Heavens, and treadeth upon the Waves of the Sea. Which maketh *Arcturus*, *Orion*, and *Pleiades*, and the *Chambers of the South*. Which doeth great Things past finding out, yea, and Wonders without Number.

x. 8--12.
Thine Hands have made me and fashioned me together round about; yet thou dost destroy me. Remember, I beseech thee, that thou hast made me as the Clay, and wilt thou bring me into Dust again? Hast Thou not poured me out as Milk, and cruddled me like Cheese? Thou hast clothed me with Skin and Flesh, and hast fenced me with Bones and Sinews. Thou hast granted me Life and Favour, and thy Visitation hath preserved my Spirit.

xii. 7--10.
But ask now the Beasts, and they shall teach thee; and the Fowls of the Air and they shall tell thee: Or speak to the Earth, and it shall teach thee; and the Fishes of the Sea shall declare unto thee. Who knoweth not in all these, that the Hand of the Lord hath wrought this? In whose Hand is the Soul of every living Thing, and the Breath of all Mankind.

xxii. 12.
Is not God in the Heighth of Heaven? And behold the Heighth of the Stars how high they are.

xxvi. 7--14.
He stretcneth out the *North* over the empty Place, and hangeth the Earth upon nothing. He bindeth up the Waters in his thick Clouds, and the Cloud is not rent under them. He holdeth back the Face of his Throne, and spreadeth his Cloud upon it. He hath compassed the Waters with Bounds, until the Day and Night come to

an end. The Pillars of Heaven tremble, and are aftonifhed at his Reproof. He divideth the Sea with his Power, and by his Underftanding he fmiteth through the Proud. By his Spirit he hath garnifhed the Heavens; his Hand hath formed the crooked Serpent. Lo, thefe are Parts of his Ways: but how little a Portion is heard of him? but the Thunder of his Power who can underftand?

God underftandeth the Way of Wifdom, and he knoweth the Place thereof. For he looketh to the Ends of the Earth, and feeth under the whole Heaven. To make the Weight for the Winds, and he weigheth the Waters by meafure. When he made a Decree for the Rain, and a Way for the Lightning of the Thunder: Then did he fee it, and declare it, he prepared it, yea and fearched it out. And unto Man he faid, behold, the Fear of the Lord that is Wifdom, and to depart from Evil is Underftanding. xxviii. 23--28.

Behold, God is great, and we know him not, neither can the Number of his Years be fearched out. For he maketh fmall the Drops of Water: They pour down Rain according to the Vapour thereof: Which the Clouds do drop, and diftill upon Man abundantly. Alfo can any underftand the Spreadings of the Clouds, or the Noife of his Tabernacle? Behold, he fpreadeth his Light upon it, and covereth the Bottom of the Sea. For by them judgeth he the People, he giveth Meat in Abundance. With Clouds he covereth the Light; and commandeth it not to fhine, by the Cloud that cometh betwixt. The Noife thereof fheweth concerning it, the Cattel alfo concerning the Vapour. xxxvi. 26-33.

At

xxxvii.
1--34.

At this also my Heart trembleth, and is moved out of his Place. Hear attentively the Noise of his Voice, and the Sound that goeth out of his Mouth. He directeth it under the whole Heaven, and his Lightning unto the Ends of the Earth. After it a Voice roareth: He thundreth with the Voice of his Excellency, and he will not stay them when his Voice is heard. God thundreth marvellously with his Voice; great Things doth he, which we cannot comprehend. For he saith to the Snow, Be thou on the Earth; likewise to the small Rain, and to the great Rain of his Strength. He sealeth up the Hand of every Man, that all Men may know his Work. Then the Beasts go into Dens, and remain in their Places. Out of the *South* cometh the Whirlwind: and Cold out of the *North*. By the Breath of God, Frost is given: and the Breadth of the Waters is straitned. Also by watering he wearieth the thick Cloud: He scattereth his bright Cloud. And it is turned round about by his Counsels: That they may do whatsoever he commandeth them upon the Face of the World in the Earth. He causeth it to come, whether for Correction, or for his Land, or for Mercy. Hearken unto this, O *Job*: Stand still and consider the wondrous Works of God. Dost thou know when God disposed them, and caused the Light of his Cloud to shine? Dost thou know the Balancings of the Clouds, the wondrous Works of him which is perfect in Knowledge? How thy Garments are warm, when he quieteth the Earth by the *South-wind*? Hast thou with him spread out the Sky, which is strong, and as a Molten Looking-Glass? Teach us what we shall say
unto

of RELIGION.

unto him ; for we cannot order our Speech by reason of Darkness. Shall it be told him that I speak? if a Man speak, surely he shall be swallowed up. And now Men see not the bright Light which is in the Clouds: But the Wind passeth and cleanseth them. Fair Weather cometh out of the North: With God is terrible Majesty. Touching the Almighty, we cannot find him out: He is excellent in Power, and in Judgment, and in Plenty of Justice: He will not afflict. Men do therefore fear him: He respecteth not any that are wise of Heart.

[*See Chap.* xxxviii, xxxix, xl, xli. Gen i. *with* 4 Esd. vi. 38--54.]

Moses.] And lest thou lift up thine Eyes unto Heaven, and when thou seest the Sun, and the Moon, and the Stars, even all the Host of Heaven, shouldest be driven to worship them, and serve them, which the Lord thy God hath divided unto all Nations under the whole Heaven. Deut. iv. 19.

Joshua.] And as soon as we had heard these Things our Hearts did melt, neither did there remain any more Courage in any Man, because of you: For the Lord your God, he is God in Heaven above, and in Earth beneath. Joshua ii. 11.

Nehemiah.] Then the *Levites, Jeshua* and *Kadmiel, Bani, Hashabniah, Sherebiah, Hodijah, Shebaniah,* and *Pethahiah*; said, Stand up and bless the Lord your God for ever and ever; and Blessed be thy glorious Name, which is exalted above all Blessing and Praise. Thou, even Thou art Lord alone, thou hast made Heaven, the Heaven of Heavens with all their Host, the Earth and all Things that are therein, the Seas and all that is therein, and thou preservest them all, and the Host of Heaven worshippeth thee. Nehemiah ix. 5, 6.

M *David.*

Astronomical Principles

Pſal. viii. 3–9.

David.] When I conſider thy Heavens, the Work of thy Fingers, the Moon and the Stars which thou haſt ordained; What is Man that thou art mindful of him? And the Son of Man, that thou viſiteſt him? For thou haſt made him a little lower than the Angels, and haſt crowned him with Glory and Honour. Thou madeſt him to have Dominion over the Works of thy Hands; thou haſt put all Things under his Feet: All Sheep and Oxen, yea, and the Beaſts of the Field: The Fowl of the Air, and the Fiſh of the Sea, and whatſoever paſſeth through the Paths of the Seas. O Lord our Lord, how excellent is thy Name in all the Earth!

xix. 1–6.

The Heavens declare the Glory of God: And the Firmament ſheweth his Handy-work. Day unto Day uttereth Speech, and Night unto Night ſheweth Knowledge. There is no Speech nor Language, where their Voice is not heard. Their Line is gone out through all the Earth, and their Words to the end of the World: In them hath he ſet a Tabernacle for the Sun. Which is as a Bridegroom coming out of his Chamber, and rejoyceth as a ſtrong Man to run a Race. His going forth is from the End of the Heaven, and his Circuit unto the Ends of it: And there is nothing hid from the Heat thereof.

cxlviii. 1––13.

Praiſe ye the Lord. Praiſe ye the Lord from the Heavens: praiſe him in the Heights. Praiſe ye him all his Angels: praiſe him all his Hoſts. Praiſe ye him Sun and Moon: praiſe him all ye Stars of Light. Praiſe him ye Heavens of Heavens, and ye Waters that be above

the

the Heavens. Let them praife the Name of the Lord: for he commanded, and they were created. He hath alfo ftablifhed them for ever and ever: He hath made a Decree which fhall not pafs. Praife the Lord from the Earth, ye Dragons, and all Deeps. Fire and Hail, Snow and Vapour, ftormy Wind fulfilling his Word. Mountains and all Hills, fruitful Trees and all Cedars. Beafts and all Cattel, creeping Things, and flying Fowl. Kings of the Earth, and all People; Princes, and all Judges of the Earth. Both young Men and Maidens, old Men and Children. Let them praife the Name of the Lord: for his Name alone is excellent, his Glory is above the Earth and Heaven.

[*See* civ, cxxxix, cxlv].

Solomon.] But will God indeed dwell on the Earth? Behold, the Heaven, and Heaven of Heavens cannot contain thee, how much lefs this Houfe that I have builded. 1 Kings viii. 27.

But who is able to build Him an Houfe, feeing the Heaven, and Heaven of Heavens cannot contain him? Who am I then that I fhould build him an Houfe, fave only to burn Sacrifice before him? 2 Chron. ii. 6.

The Lord by Wifdom hath founded the Earth; by Underftanding hath he eftablifhed the Heavens. By his Knowledge the Depths are broken up, and the Clouds drop down the Dew. Prov. iii. 19, 20.

As thou knoweft not what is the Way of the Spirit, nor how the Bones do grow in the Womb of her that is with Child. even fo thou knoweft not the Works of God who maketh all. Ecclef. xi. 5.

Ifaiah.]

Isaiah xl. 12--17.

Isaiah.] Who hath measured the Waters in the Hollow of his Hand? And meted out Heaven with the Span, and comprehended the Dust of the Earth in a Measure, and weighed the Mountains in Scales, and the Hills in a Balance. Who hath directed the Spirit of the Lord, or being his Counsellor hath taught him? With whom took he Counsel, and who instructed him, and taught him in the Path of Judgment, and taught him Knowledge, and shewed to him the Way of Understanding? Behold, the Nations are as a Drop of a Bucket, and are counted as the small Dust of the Balance: Behold, he taketh up the Isles as a very little Thing. And *Lebanon* is not sufficient to Burn, nor the Beasts thereof sufficient for a Burnt Offering. All Nations before him are as nothing, and they are counted to him less than Nothing, and Vanity.

xlv. 5--8,

I am the Lord, and there is none else, there is no God besides me: I girded thee, though thou hast not known me: That they may know from the rising of the Sun, and from the *West*, that there is none besides me, I am the Lord, and there is none else. I form the Light, and create Darkness: I make Peace, and create Evil: I the Lord do all these Things. Drop down, ye Heavens, from above, and let the Skies pour down Righteousness! Let the Earth open, and let them bring forth Salvation, and let Righteousness spring up together: I the Lord have

ver. 18.

created it.---For thus saith the Lord that created the Heavens, God himself that formed the Earth and made it, he hath established it, he created it not in vain, he formed it to be inhabited, I am the Lord, and there is none else.

Thus

of Religion.

Thus saith the Lord, the Heaven is my Throne, and the Earth is my Footstool: Where is the House that ye build unto me? And where is the Place of my Rest? For all those Things hath mine Hand made, and all those Things have been, saith the Lord: But to this Man will I look, even to him that is poor, and of a contrite Spirit, and trembleth at my Word. *lxvi. 1, 2*

Jeremiah.] Fear ye not me? Saith the Lord: Will ye not tremble at my Presence; which have placed the Sand for the Bound of the Sea, by a perpetual Decree that it cannot pass it; and though the Waves thereof toss themselves, yet can they not prevail; though they roar, yet can they not pass over it? *Jeremiah v. 22.*

But the Lord is the true God, he is the living God, and an everlasting King: At his Wrath the Earth shall tremble, and the Nations shall not be able to abide his Indignation. Thus shall ye say unto them, the Gods that have not made the Heavens, and the Earth, even they shall perish from the Earth, and from under these Heavens. He hath made the Earth by his Power, he hath established the World by his Wisdom, and hath stretched out the Heavens by his Discretion. When he uttereth his Voice, there is a Multitude of Waters in the Heavens, and he causeth the Vapours to ascend from the Ends of the Earth: He maketh Lightnings with Rain, and bringeth forth the Wind out of his Treasures. *x. 10--13.*

Am I a God at hand, saith the Lord, and not a God afar off? Can any hide himself in secret Places that I shall not see him? saith the Lord: Do not I fill Heaven and Earth? saith the Lord. *xxiii. 23, 24.*

M 3 Thus

xxxi. 35.   Thus saith the Lord, which giveth the Sun for a Light by Day, and the Ordinances of the Moon and of the Stars for a Light by Night, which divideth the Sea when the Waves thereof roar; the Lord of Hosts is his Name.

xxxii. 17.   Ah Lord God, Behold thou haſt made the Heaven and the Earth, by thy great Power and ſtretched-out Arm, and there is nothing too hard for thee.

Daniel iv. 34, 35.   *Daniel.*] And at the End of the Days, I *Nebuchadnezzar* lift up mine Eyes unto Heaven, and mine Underſtanding returned unto me, and I bleſſed the moſt High, and I praiſed and honoured him that liveth for ever, whoſe Dominion is an Everlaſting Dominion, and his Kingdom is from Generation to Generation. And all the Inhabitants of the Earth are reputed as nothing: And he doth according to his Will in the Army of Heaven, and among the Inhabitants of the Earth: And none can ſtay his Hand, or ſay unto him, What doſt thou?

v. 22, 23.   And thou his Son, O *Belſhazzar*, haſt not humbled thine Heart, though thou kneweſt all this: But haſt lifted up thy ſelf againſt the Lord of Heaven, and they have brought the Veſſels of his Houſe before thee, and thou, and thy Lords, thy Wives and thy Concubines have drunk Wine in them, and thou haſt praiſed the Gods of Silver and Gold, of Braſs, Iron, Wood and Stone, which ſee not, nor hear, nor know. And the God in whoſe Hand thy Breath is, and whoſe are all thy Ways, haſt thou not glorified.

Amos v. 8.   *Amos.*] Seek him that maketh the Seven Stars and *Orion*, and turneth the Shadow of Death into the Morning, and maketh the Day dark with

of RELIGION.

with Night: That calleth for the Waters of the Sea, and poureth them out upon the Face of the Earth, the Lord is his Name.

Jonah.] And he said unto them, I am an Hebrew, and I fear the Lord the God of Heaven, which hath made the Sea and the Dry Land. — Jonah i. 9.

Habakkuk.] A Prayer of *Habakkuk* the Prophet upon *Sigionoth*. O Lord, I have heard thy Speech and was afraid: O Lord, revive thy Work in the midst of the Years, in the midst of the Years make known; in Wrath remember Mercy. God came from *Teman*, and the Holy One from Mount *Paran*. *Selah*. His Glory covered the Heavens, and the Earth was full of his Praise. And his Brightness was as the Light, he had Horns coming out of his Hand, and there was the hiding of his Power. Before him went the Pestilence, and burning Coals went forth at his Feet. He stood and measured the Earth: He beheld, and drove asunder the Nations, and the everlasting Mountains were scattered, the perpetual Hills did bow: His Ways are everlasting. I saw the Tents of *Cushan* in Affliction: And the Curtains of the Land of *Midian* did tremble. Was the Lord displeased against the Rivers? Was thine Anger against the Rivers? Was thy Wrath against the Sea, that thou didst ride upon thine Horses, and thy Chariots of Salvation? Thy Bow was made quite naked, according to the Oaths of the Tribes, even thy Word. *Selah*. Thou didst cleave the Earth with Rivers. The Mountains saw thee, and they trembled: The overflowing of the Water passed by: The Deep uttered his Voice, and lift up his Hands on high. The Sun and Moon stood still in their Habitation: At the — Habakkuk iii. 1--19.

Light

Light of thine Arrows they went, and at the shining of thy glittering Spear. Thou didst march through the Land in Indignation, thou didst thresh the Heathen in Anger. Thou wentest forth for the Salvation of thy People, even for Salvation with thine Anointed; thou woundedst the Head out of the House of the Wicked, by discovering the Foundation unto the Neck. *Selah.* Thou didst strike through with his Staves the Head of his Villages: They came out as a Whirlwind to scatter me: Their Rejoycing was as to devour the poor secretly. Thou didst walk through the Sea with thine Horses, through the Heap of great Waters. When I heard, my Belly trembled: My Lips quivered at the Voice: Rottenness entred into my Bones, and I trembled in my self, that I might rest in the Day of Trouble: When he cometh up unto the People, he will invade them with his Troops. Although the Fig-tree shall not blossom, neither shall Fruit be in the Vines, the Labour of the Olive shall fail, and the Fields shall yield no Meat, the Flock shall be cut off from the Fold, and there shall be no Herd in the Stalls: Yet I will rejoyce in the Lord, I will joy in the God of my Salvation. The Lord God is my Strength, and he will make my Feet like Hinds Feet, and he will make me to walk upon mine high Places. To the chief Singer on my stringed Instruments.

Eccl. xlii. 15:-25.

Sirach.] I will now remember the Works of the Lord, and declare the Things that I have seen: In the Words of the Lord are his Works. The Sun that giveth Light, looketh upon all Things, and the Work thereof is full of the Glory of the Lord. The Lord hath not given

Power to the Saints to declare all his marvellous Works, which the Almighty Lord firmly settled, that whatsoever is might be established for his Glory. He seeketh out the Deep, and the Heart, and considereth their crafty Devices: For the Lord knoweth all that may be known, and he beholdeth the Signs of the World. He declareth the Things that are past, and for to come, and revealeth the Steps of hidden Things. No Thought escapeth him, neither any Word is hidden from him. He hath garnished the excellent Works of his Wisdom, and he is from Everlasting to Everlasting: Unto him may nothing be added, neither can he be diminished, and he hath no need of any Counsellor. Oh how desirable are all his Works! And that a Man may see even to a Spark. All these Things live and remain for ever, for all Uses, and they are all Obedient. All Things are double one against another: And He hath made nothing imperfect. One Thing establisheth the good of another: And who shall be filled with beholding his Glory?

The Pride of the Height, the clear Firmament, the Beauty of Heaven, with his glorious Shew; The Sun when it appeareth, declaring at his Rising a marvellous Instrument, the Work of the most High. At Noon it parcheth the Country, and who can abide the burning Heat thereof? A Man blowing a Furnace is in Works of Heat, but the Sun burneth the Mountains three Times more; breathing out fiery Vapours, and sending forth bright Beams, it dimmeth the Eyes. Great is the Lord that made it, and at his Commandment it runneth hastily. He made the Moon also to serve in her Season, for a Declaration

xliii. 1-33.

claration of Times, and a Sign of the World. From the Moon is the sign of Feasts, a Light that decreaseth in her Perfection. The Month is called after her Name, increasing wonderfully in her changing, being an Instrument of the Armies above, shining in the Firmament of Heaven; The Beauty of Heaven, the Glory of the Stars, an Ornament giving Light in the highest Places of the Lord. At the Commandment of the Holy One, they will stand in their Order, and never faint in their Watches. Look upon the Rainbow, and praise him that made it, very beautiful it is in the Brightness thereof. It compasseth the Heaven about with a glorious Circle, and the Hands of the most High have bended it. By his Commandment he maketh the Snow to fall apace, and sendeth swiftly the Lightnings of his Judgment. Through this the Treasures are opened, and Clouds fly forth as Fowls. By his great Power he maketh the Clouds firm, and the Hailstones are broken small. At his Sight the Mountains are shaken, and at his Will the South-wind bloweth. The noise of the Thunder maketh the Earth to tremble; so doth the Northern Storm and the Whirlwind: As Birds flying he scattereth the Snow, and the falling down thereof is as the Lighting of Grashoppers. The Eye marvelleth at the Beauty of the Whiteness thereof, and the Heart is astonished at the raining of it. The Hoar-frost also as Salt he poureth on the Earth, and being congealed, it lieth on the top of sharp Stakes. When the cold North-wind bloweth, and the Water is congealed into Ice, it abideth upon every gathering together of Water, and cloatheth the Water as

with

with a Breast-plate. It devoureth the Mountains, and burneth the Wilderness, and consumeth the Grass as Fire. A present Remedy of all is a Mist *coming speedily*: A Dew coming after Heat, refresheth. By his Counsel he appeaseth the Deep, and planteth Islands therein. They that sail on the Sea, tell of the Danger thereof, and when we hear it with our Ears, we marvel thereat. For therein be strange and wondrous Works, Variety of all Kinds of Beasts, and Whales created. By him the end of them hath prosperous Success, and by his Word all Things consist. We may speak much, and yet come short: Wherefore in sum, he is all. How shall we be able to magnify him? For he is great above all his Works. The Lord is terrible, and very great, and marvellous is his Power. When ye glorify the Lord, exalt him as much as ye can; for even yet will he far exceed: And when ye exalt him, put forth all your Strength, and be not weary; for ye can never go far enough. Who hath seen him that he might tell us? and who can magnify him as he is? There are yet hid greater Things then these be, for we have seen but a few of his Works. For the Lord hath made all Things, and to the Godly hath he given Wisdom.

Baruch.] Who hath gone up into Heaven and taken Wisdom, and brought her down from the Clouds? Who hath gone over the Sea, and found her, and will bring her for pure Gold? No Man knoweth her Way, nor thinketh of her Path. But he that knoweth all Things, knoweth her, and hath found her out with his Understanding: He that prepared the Earth for ever-

Baruch iii. 29--35.

evermore, hath filled it with four-footed Beasts. He that sendeth forth Light, and it goeth; calleth it again, and it obeyeth him with fear. The Stars shined in their Watches, and rejoyced: When he calleth them, they say, Here we be: and so with chearfulness they shewed Light unto him that made them. This is our God, and there shall none other be accounted of in comparison of him.

[*See Song of the Three Children at large.*]

Manasses Prayer.

Manasses.] O Lord, Almighty God of our Fathers, *Abraham, Isaac,* and *Jacob,* and of their Righteous Seed, who hast made Heaven and Earth, with all the Ornament thereof; who hast bound the Sea by the Word of thy Commandment; who hast shut up the deep, and sealed it by thy terrible and glorious Name; whom all Men fear, and tremble before thy Power; for the Majesty of thy Glory cannot be born, and thine angry threatning towards Sinners is importable: but thy merciful Promise is unmeasurable, and unsearchable: for thou art the most High Lord, of great Compassion, Long-suffering, very Merciful, and repentest of the evils of Men.

Acts xv.i. 23--28.

Paul.] For as I passed by, and beheld your Devotions, I found an Altar with this Inscription, TO THE UNKNOWN GOD. Whom therefore ye ignorantly Worship, him declare I unto you. God that made the World, and all Things therein, seeing that he is Lord of Heaven and Earth, dwelleth not in Temples made with Hands: Neither is worshipped with Mens Hands, as though he needed any Thing, seeing he giveth to all Life and Breath, and all Things; And hath made of one Blood all Nations

tions of Men, for to dwell on all the Face of the Earth: And hath determined the Times before appointed, and the Bounds of their Habitation. That they should seek the Lord, if haply they might feel after him, and find him, though he be not far from every one of us. For in him we Live, and Move, and have our Being; as certain also of your own Poets have said, for we are also his Offspring.

Because that which may be known of God, is manifest in them; for God hath shewed it unto them. For the Invisible Things of him are clearly seen from the Creation of the World, being understood by the Things that are made, *even* his Eternal Power and Godhead; so that they are without Excuse. *Rom. i. 19, 20.*

John.] And every Creature which is in Heaven, and on the Earth, and under the Earth, and such as are in the Sea, and all that are in them, heard I, saying, Blessing, and Honour, and Glory, and Power, *be* unto him that sitteth upon the Throne, and unto the Lamb for ever and ever. *Rev. v. 13.*

Clement.] The Heavens holding fast to his Appointment, are subject to him in Peace. Day and Night accomplish the Courses that he has allotted unto them, not disturbing one another. The Sun and Moon, and all the several Companies and Constellations of the Stars, run the Courses that he has appointed to them in Concord, without departing in the least from them. The Fruitful Earth yields its Food plentifully in due Season both to Man and Beast, and to all that is upon it, according to his *1 Epistle, § 20.*

his Will; not disputing, nor altering any Thing of what was order'd by him. So also the untrodden and unsearchable Floods of the Deep are kept in by his Command: And the Conflux of the vast Sea being brought together at the Creation into its several Collections, passes not the Bounds that he has set to it; but as he then appointed it, so it remains. For he said, *Hitherto shalt thou come, and thy Floods shall be broken within thee.* The Ocean, unpassable to Mankind, and the Worlds that are beyond it, are govern'd by the same Commands of their Master. Spring and Summer, Autumn and Winter, give Place peaceably to each other. The several Quarters of the Winds, fulfil their Work in their Seasons, without offending one another. The ever-flowing Fountains, made both for Pleasure and Health, never fail to reach out their Breasts to support the Life of Men. Even the smallest Creatures live together in Peace and Concord with each other. All these has the Great Creator and Lord of all, commanded to observe Peace and Concord; being Good to all: But especially to Us who flee to his Mercy through our Lord Jesus Christ, to whom be Glory and Majesty for Ever and Ever. *Amen.*

vii. 34. *Apostles in their Constitutions.*] Thou art Blessed, O Lord, the King of Ages, who by Christ hast made the whole World, and by him in the Beginning didst reduce into order the disorder'd Parts. Who dividedst the Waters from the Waters by a Firmament; and didst put into them a Spirit of Life; who didst fix the Earth, and stretch out the Heaven, and didst dispose every Creature by an accurate Constitution: For by thy Power,

Power, O Lord, the World is Beautify'd, the Heaven is fix'd as an Arch over us, and is rendred illustrious with Stars for our Comfort in the Darkness: The Light also and the Sun were begotten for Days, and the Production of Fruit; and the Moon for the Change of Seasons, by its Increase and Diminutions; and one was called Night, and the other Day. And the Firmament was exhibited in the midst of the Abyss, and thou commandest the Waters to be gathered together, and the dry Land to appear. But as for the Sea it self, who can possibly describe it? Which comes with Fury from the Ocean, yet runs back again, being stopp'd by the Sand at thy Command; for thou hast said, * *Thereby shall her Waves be broken.* Thou hast also made it capable of supporting little and great Creatures, and made it Navigable for Ships. Then did the Earth become Green, and was planted with all sorts of Flowers, and the Variety of several Trees; and the shining Luminaries, the Nourishers of those Plants, preserve their unchangeable Course, and in nothing depart from thy Command. But where thou biddest them, there do they rise and set, for Signs of the Seasons, and of the Years, making a constant Return of the Work of Men. Afterwards the Kinds of the several Animals were created, those belonging to the Land, to the Water, to the Air, and both to Air and Water; and the Artificial Wisdom of thy Providence does still impart to every one a suitable Providence. For as he was not unable to produce different Kinds, so neither has he disdain'd to exercise a different Providence towards every one. And

* Job xxxviii. 11.

at the Conclusion of the Creation thou gavest Direction to thy Wisdom, and formedst a reasonable Creature, as the Citizen of the World, saying, *Let us make Man after our Image, and after our Likeness*; and hast exhibited him as the Ornament of the World, and formed him a Body out of the Four Elements, those primary Bodies, but hadst prepared a Soul out of nothing, and bestowedst upon him his Five Senses, and didst set over his Sensations a Mind, as the Conducter of the Soul. And, besides all these Things, O Lord God, who can worthily declare the Motion of the Rainy Clouds, the shining of the Lightning, the Noise of the Thunder, in order to the Supply of proper Food, and the most agreeable Temperature of the Air? But when Man was disobedient, thou didst deprive him of the Life which should have been his Reward; yet didst thou not destroy him for ever, but laidst him to Sleep for a Time, and thou didst by Oath call him to a Resurrection, and loosedst the Bond of Death; O thou Reviver of the Dead, through Jesus Christ, who is our Hope.

Gen. i. 25.

vii. 35.

Great art thou, O Lord Almighty, and Great is thy Power, and of thy Understanding there is no Number. Our Creator and Saviour, rich in Benefits, Long-suffering, and the Bestower of Mercy, who dost not take away thy Salvation from thy Creatures; for thou art good by Nature, and sparest Sinners, and invitest them to Repentance; for Admonition is the Effect of thy Bowels of Compassion; for how should we abide if we were requir'd to come

come to Judgment immediately, when after so much Long-suffering, we hardly get clear of our miserable Condition? The Heavens declare thy Dominion, and the Earth shakes with Earthquakes, and, hanging upon nothing, declares thy unshaken Stedfastness. The Sea raging with Waves, and feeding a Flock of Ten thousand Creatures, is bounded with Sand, as standing in awe at thy Command; and compels all Men to cry out, * *How great are thy Works, O Lord! In Wisdom hast thou made them all! The Earth is full of thy Creation.* And the bright Host of Angels, and the Intellectual Spirits say to *Palmoni*, † *There is but one Holy Being:* And the Holy Seraphim, together with the Six-winged Cherubim, who sing to Thee their Triumphal Song, cry out with never-ceasing Voices, * *Holy, Holy, Holy, Lord God of Hosts; Heaven and Earth are full of thy Glory:* And the other Multitudes of the Orders, Angels, Arch-Angels, Thrones, Dominions, Principalities, Authorities and Powers, cry aloud, and say, ‖ *Blessed be the Glory of the Lord out of his Place.* But *Israel*, thy Church on Earth, taken out of the *Gentiles*, emulating the Heavenly Powers, Night and Day, with a full Heart, and a willing Soul, sings, * *The Chariot of God is ten thousandfold, thousands of the prosperous: The Lord is among them in* Sinai, *in the holy Place.* The Heaven knows him who fix'd it as a Cube of Stone, in the Form of an Arch, upon nothing; who united the Land and Water to one another, and scatter'd the Vital Air all abroad, and conjoin'd Fire therewith for Warmth, and Comfort against Darkness. The Choir of Stars strikes us with Admiration,

* Psal. ciii. 24.

† Dan. viii. 13.

* Isa. vi. 3.

‖ Ezek. iii. 12.

* Psal. lxvii. 18.

Admiration, declaring him that numbers them, and shewing him that names them; the Animals declare him that puts Life into them; the Trees shew him that makes them grow: All which Creatures, being made by thy Word, shew forth the Greatness of thy Power. Wherefore every Man ought to send up an Hymn from his very Soul to thee, through Christ, in the Name of all the rest, since he has Power over them all, by thy Appointment. For thou art kind in thy Benefits, and beneficent in thy Bowels of Compassion; who alone art Almighty; for when thou willest, to be able is present with Thee; for thy eternal Power both quenches Flame, and stops the Mouths of Lions, and tames Whales, and raises up the Sick, and over-rules the Power of all Things, and overturns the Host of Enemies, and casts down a People numbred in their Arrogance. Thou art he who art in Heaven, he who art on Earth, he who art in the Sea, he who art in finite Things, thy Self unconfin'd by any thing: For of thy Majesty there is no Boundary: For 'tis not ours, O Lord, but the Oracle of thy Servant, who said, † *And thou shalt know in thine Heart, that the Lord thy God he is God, in Heaven above, and on Earth beneath, and there is none besides Thee*: For there is no God besides Thee alone, there is none holy besides Thee, the Lord, the God of Knowledge, the God of Saints, holy above all holy Beings; for they are sanctified by thy Hands: Thou art Glorious, and highly exalted, invisible by Nature, and unsearchable in thy Judgments; whose Life is without Want, whose Duration can never fail, whose Operation is without Toil, whose

† Deut. iv. 39.

whose Greatness is unlimited, whose Excellency is perpetual, whose Habitation is inaccessible, whose Dwelling is unchangeable, whose Knowledge is without Beginning, whose Truth is immutable, whose Work is without Assistants, whose Dominion cannot be taken away, whose Monarchy is without Succession, whose Kingdom is without End, whose Strength is irresistible, whose Army is very numerous. For thou art the Father of Wisdom, the Creator of the Creation, by a Mediator, as the Cause. The Bestower of Providence, the Giver of Laws, the Supplier of Want, the Punisher of the Wicked, and the Rewarder of the Righteous; the God and Father of Christ, and the Lord of those that are Pious towards Him; whose Promise is infallible, whose Judgment without Bribes, whose Sentiments are immutable, whose Piety is incessant, whose Thanksgiving is everlasting, through whom Adoration is worthily due to Thee from every rational and holy Nature.

" *Hermas.*] Behold the mighty Lord, who by his [Vis. i. 3]
" invincible Power, and with his excellent Wis-
" dom made the World, and by his glorious
" Counsel encompassed the Beauty of his Crea-
" ture, and with the Word of his Strength fix'd
" the Heaven, and founded the Earth upon the
" Waters; and by his powerful Vertue esta-
" blish'd his Holy Church, which he hath
" blessed: Behold, he will remove the Heavens,
" and the Mountains, the Hills and the Seas;
" and all things shall be made Plain for his
" Elect; that he may render unto them the Pro-
" mise which he has promised with much Ho-
" nour and Joy; if so be that they shall keep
" the

  " the Commandments of God, which they have
  " received with great Faith.

viii. 20. *Author of the Recognitions.*] But somebody may say, that these things are done by *Nature*. Now in this case the Contention is only about a Word. For while 'tis certain that the World is the Work of a Mind, and of Reason, what you call *Nature* I call *God the Creator*. And clear it is, that neither formerly nor now could it be, that either the Species of Bodies, adorn'd with such necessary Distinctions; or the Faculties of the Mind, should be made by any Labour, without Reason, and without Sense. And now, if you look on the Philosophers as proper Witnesses in this case, *Plato* gives us his Testimony in his *Timæus*; where in his discussion of this Question, about the Frame of the World, Whether it always was, or had a Beginning, he pronounces that it was made. For, says he, 'tis visible, palpable, and corporeal; and all things of that Nature were certainly made. Now what was made, has without question some Author by whom it was made. But then, as he adds, To discover this Maker and Parent of all Things, is no easy thing; and when you have discover'd him, to impart your discovery to the vulgar, is plainly impossible. These are certainly *Plato*'s words. But supposing that he and the other Philosophers among the *Greeks* had been dispos'd to say nothing about the making of the World, would it not still be a plain case to all that had common Understanding? For what Man is there, I mean one of at least some small Capacity, who upon the sight of an House with all its Furniture fitted for Mens various Necessities, whose Top is
            adorn'd

adorn'd with a spherical Cupolo, beautify'd with variety of splendid Draughts and several sorts of Pictures, and adorn'd with the fairest and largest Lights; who is there, I say, that upon the view of such a Fabrick will not immediately pronounce that it was fram'd by a most wise and most powerful Architect? And can any one be found so foolish, as upon the sight of the Work of Heaven, and the view of the splendor of the Sun and Moon, the regular Course of the Stars, with their various Kinds and Motions, and that sees all determin'd by proper Laws, and to suitable Periods; to forbear to cry out, that these things were made by a wise and rational Artificer, or rather by Wisdom and Reason it self?

But now if you desire to be a Follower of others of the *Greek* Philosophers, and are vers'd in Mechanicks, what they deliver about these Celestial things must have certainly come to your knowledge: For they suppose that the Heavens are like a Sphere, on every side evenly situate, and having the same respect to every part, and equally distant from the Center of the Earth; and that therefore they stand so firm by the equality of their Libration, that the evenness of their Situation does not allow them to bend any one way more than another; and that by this means the Sphere is sustain'd without any Prop to support it. Now if this Machine of the World bears really this Similitude, there is a clear demonstration of Divine Workmanship therein. But if, as others suppose, this spherical Arch is supported by the Waters; either as it floats on their Surface, or as it turns round within them, even on those

viii. 21.

Hypotheses the Workmanship of the great Artificer is manifested therein.

viii. 22.

But lest the Arguments of this kind, which all are not capable to understand, should seem of an uncertain nature, let us proceed to such as every one can comprehend. Who is it that has order'd the Courses of the Stars with so great Judgment, and appointed their times of Rising and Setting, and ordain'd every one of them to hold its course in the Heavens in certain and fix'd Periods? Who is it that has permitted some of them to go always Westward, and others to return sometimes Eastward? Who is it that has fix'd Limits to the Courses of the Sun, that he might determine Hours, and Days, and Months, and the Vicissitudes of Seasons by its different Motions; and by the sure adjustment of its Course distinguish those Seasons into Winter first, then the Spring, after that the Summer, and Autumn; so as still to determine the annual Period by the same Revolutions? Who is there, I say, but must pronounce the Divine Wisdom it self to be the Manager of so regular a System? And so much for that Hypothesis which the *Greeks* have form'd about the System of the Heavens.

viii. 23.

What also can be said to those Appearances which belong to the Land and to the Sea? Are not we plainly taught by them, that God did not only make these Parts of the World, but that he exercises a Providence over them also? For therefore are there high Mountains in some certain Places in every part of the World, that the Air which is, as it were, compress'd and straiten'd by them, may, according to the appointment of God, be crowded and forc'd out

for

for Winds; whereby the Fruits grow, and the Heat of Summer is temper'd, at the time when the warm *Pleiades* are heated by the fierceness of the Sun. But you will say, why was there such an intense Heat in the Sun at all, which should require to be temper'd? Pray how could the Fruits of the Earth, which are so necessary for the Uses of Mankind, be ripen'd without it? Besides, take notice of another Thing, that near the *Equator*, where the greatest Heat is, there is no great compression of the Clouds; nor does any mighty quantity of Rain fall there, left it should breed Diseases among the Inhabitants. For moist Clouds, if they be as it were bak'd with an intense Heat, do render the Air corrupt and pestilential. As also the Earth, when it receives over-warm Rain, does not afford Nourishment to the Corn, but destroys it. Which Management who can doubt but 'tis the Effect of the Divine Providence? To conclude with the Case of *Egypt*, which because 'tis near the burning Heat of *Ethiopia*, and so if it stood in absolute Necessity of Rain, would have its Air intolerably corrupted; its Fields are therefore supply'd, not by Rain, which is deriv'd from the Clouds; but they enjoy a kind of terrestrial Rain, by the Inundations of the *Nile*.

What is also to be said about the Fountains and Rivers, which run with a constant Current into the Sea? and yet all is so fitted by the Divine Providence, that those Rivers do not want a plentiful Current of Water; and yet that the Sea, which receives such vast Quantities of Water, seems not to be augmented; but those Elements continue in the same Proportion, both those which carry, and those which receive that Supply

viii. 24.

Supply of Water; the salt Water still naturally consuming the sweet Streams mixed with it. Herein therefore the Effects of Providence are manifest, that it should make that Element Salt whereto the Course of all those Waters which it had afforded for the Uses of Men, was to carry them; that so the full Cavity of the Sea might never, in all the Series of Ages, bring upon the Earth and upon Men, any fatal Inundation of Waters. And there is no Man so foolish to suppose so great an Instance of Reason and of Providence could be taken care of by any irrational Nature.

viii. 25. What shall I say about Plants and about Animals? Is it not the Effect of Providence, that when they are to be dissolv'd by Old Age, the Plants should be repair'd again, either by young Plants, or by Seeds, which proceed from themselves; and the Animals by the Propagation of Posterity? And indeed, 'tis by the surprizing Conduct of Providence, that Milk is provided in the Breasts against the time when the Young one is born; and that the same young one, as soon as 'tis born, without any Instructor, knows where to look for the Places wherein its Nourishment is laid up for it. And then Males as well as Females are brought forth; that by the means of both, Posterity may be provided for: But now lest, as Men are ready to imagine, these Events might seem to happen according to some fix'd Course of Nature, and not by the Dispensation of the Creator, he ordain'd that some few Creatures should propagate their Kind upon Earth after a different manner, for an Indication and Sign of his Providence: That for Example, the Raven should bring forth her Young

Young at the Mouth, and the Weezle propagate at the Ear; that some sort of Fowls, as Hens, should bring forth Eggs, addle either by the Wind or the Dust; that some other Creatures should change the Male by Turns, into the Female, and every Year alter their Sex; as Hares and the Hyænæ, which they call Monsters; that some should arise out of the Earth, and take thence their Flesh, as Moles; others out of Ashes, as Vipers; others out of putrefy'd Flesh, as Wasps out of the Flesh of Horses, and Bees out of that of Kine; others out of Cows Dung, as Beetles; others out of Herbs, as the Scorpion out of *Basil*; and on the contrary, that Herbs should spring out of Animals, as Smallage and Asparagus out of the Horn of a Stag or of a Roe-Buck.

And indeed to what Purpose should I reckon up more Examples wherein the Divine Providence, by changing that Course which is suppos'd to be appointed by Nature, has in many Respects varied the Circumstances of the Birth of Animals? Whereby might be shew'd not any irrational Course of Things, but God the Disposer of all Things might rationally be demonstrated. Is there not also in another Instance, a compleat Demonstration of the Workmanship of Divine Providence? In that I mean, when Seeds that are sown are repair'd for the Uses of Human Life? Which Seeds when they are committed to the Ground, the Soil, by the Will of God, affords them that Moisture it has receiv'd, as if it were Milk for their Nourishment. For there is in the Waters a certain Power of the Spirit of God, which was afforded them at the beginning, by whose Efficiency the entire

viii. 26.

entire future Body begins to be form'd in the very Seed, and to be reftor'd again by its Stem and its Ear: For when a Grain of the Seeds is fwell'd by the Moifture, that Power of the Spirit which was beftow'd on the Waters, and being incorporeal eafily runs through certain narrow Channels of the Veins, invigorates the Seeds till they grow larger, and frames the Species of them as they grow. It comes to pafs therefore, that by the Means of the moift Element, wherein this vital Spirit is ever inferted and implanted, that not only the Corn is repair'd in general, but that it returns again, as to its Species and Form, entirely like thofe Seeds which were fow'd. Which Regularity of Operation, who that has the leaft Senfe can believe to be thus perform'd by an irrational Nature, and not by the Divine Wifdom? To conclude, Even thefe Things are form'd after the fimilitude of a Human Birth; for the Earth appears to retain the Place of the Womb, where the Seed when it is caft into it, is form'd and nourifh'd by the Power of Water and of the Spirit, as we have faid already.

viii. 27. ¹ Moreover, the Divine Providence is herein alfo to be admir'd, that it has order'd all, fo that we can indeed fee and know what is made; but how, and after what manner it is made, is hidden and conceal'd from us; that they may not be difcoverable by thofe that are unworthy, but may be difclos'd to fuch as are worthy and faithful, and have merited fuch a Favour. Now that we may prove by Experiments and Inftances, that the Seeds do not receive any Part of the Terrene Matter, but are made up entirely of the Element of Water, and of the Virtue of that

that Spirit which is included therein; Do you suppose, for Example's sake, that the Weight of an Hundred Talents of Earth were put into a large Vessel; and the several Kinds of Seeds were sow'd therein, either of Herbs, or of larger Plants; and that they had a sufficient Quantity of Water to moisten them; and let this Process be continued several Years. And then let the Grain which has sprung from them, suppose of Wheat or Barley, or of any other sort, be gather'd together, every Year's Product by it self, till the Heap of every Kind of Grain is arisen to the Weight of an Hundred Talents. Then let the Trees themselves be pluck'd up, and weigh'd; and when they are all taken out of the Vessel, yet will the Earth it self, when it is weigh'd, afford you its entire original Hundred Talents again notwithstanding. Whence then shall we say that all that Weight, and all that Quantity of different Sorts of Grain, and of the Trees has arisen? Is it not plain that 'tis from the Water? For the Earth retains its own entirely, while the Water which was poured on every one of them, wholly disappears; and all this by the powerful Efficacy of the Divine Disposal of the Creator, which by the very Element of Water both repairs the Substances, and frames the Species of such Seeds and Plants, and preserves their Species with great Increase.

From all which Instances I think it is abundantly evident to all Men, that all Things are made, and every Thing does subsist by the Skill of a wise Being, and not by the Operation of Brutal Nature. But now let us proceed, if you please, to our own Constitution, or that of

viii. 28.

of a Man, who is a little World included in the other: And let us consider with what Art he is compounded, and thereby you will see in an especial Manner the Wisdom of the Creator. Now though he be made up of different Substances, of that which is Mortal, and that which is Immortal, yet by the Skill and Providence of his Creator is it brought to pass that these different Substances, which are so widely remote one from the other, admit of an Union; For one Part is taken from the Earth, and fram'd by the Creator; while the other is deriv'd from the Immortal Substances. And yet is the Advantage of Immortality not at all infring'd by such a Conjunction. Nor is he made up of rational, and concupiscible, and irascible Parts; but such sorts of Faculties as those, are rather to be suppos'd Affections belonging to him; whereby he may be carried to those several sorts of Objects. For the Body, which consists of Bones and Flesh, owes its Original to the Seed of the Male, which Heat fetches out of the Marrow, and consigns over to the Womb, as to a proper Soil, whereto it adheres: And when it has by little and little been moisten'd, by the flowing of the Blood to it, it becomes Flesh and Bones; and is made up after the Species of him who cast in the Seed.

viii. 29.

Behold now the Contrivance of the Artificer herein! how he has inserted the Bones as certain Pillars, whereby the Flesh might be sustain'd and supported. Besides this, consider how a just Measure is preserv'd on both Sides; I mean on the right Side, and

on

on the Left; so that one Foot agrees with the other, and one Hand with the other, one Set of Fingers with the other; that so every one of them might agree with his Fellow, without the least Inequality: Which is the Case also as to the one Eye with the other, and the one Ear with the other; which Members do not only resemble and agree with one another, but are also so fram'd as to serve for the necessary Occasions of Life. The Hand, for Instance, is so dispos'd as to be fit for Work, the Feet for walking, the Eyes for seeing, as guarded by the Eye-brows; the Ears are so fram'd for Hearing, that like a Drum they send the rebounding Sound of Words deep into the Head, and even as far as the Sensation of the Soul; as does the Tongue, when 'tis mov'd upon the Teeth, supply the Place of a Quill. Those Teeth also are so form'd, that some chew and divide the Food, and send it to others, who are more inward; and those Teeth that are more inward are so fram'd, that, like Milstones, they chew and break it small; that so it may be deliver'd to the Stomach in a state fit for Digestion. Whence it is that these Teeth have the Name of *Grinders* bestowed on them.

Besides these, the Nostrils were made for the Passage of the Breath to and fro, in Expiration and Inspiration; that the natural Heat which is in the Heart, may by the Access of fresh Air, be heated or cooled by the Operation of the Lungs; which are therefore plac'd in the Breast, that by its Softness it may cherish and enliven the Heart, in whose vigo-

viii. 30.

vigorous State Life seems to consist. I say the Life, not the Soul. For what shall I say of the Substance of the Blood? Which is like a River, proceeding from a Fountain, which at first is carry'd along one Channel, but then is deriv'd farther by innumerable Veins, as by so many Pipes; and so waters the entire Soil of a human Body with vital Streams; whereto the Liver is also assisting, which is situate on the Right Side, for the more effectual Digestion of the Food, and its Conversion into Blood: While the Place of the Spleen is on the Left Side, that it may attract to it self, and after a sort cleanse the Blood of its Impurities.

viii. 31.
And as to the Contrivance of the Intestines, how wonderful is it! For therefore are they join'd together in long Foldings, like Circles, that they may leisurely throw off the Remains of the Food after Digestion, that so the Receptacles of the Nourishment may not be suddenly empty'd; and yet there may no Hindrance arise from the Food that is taken afterwards. But therefore are they contriv'd to be Membranaceous, that the Parts without them may by degrees receive from them their moist Nutriment; that so it may not go away at once, and leave the Bowels themselves empty; nor be hindred by the Thickness of the Skin, and leave the other Parts dry, and disorder thereby the whole Human Fabrick with inevitable Thirst.

viii. 32.
Moreover, who is there whom the Position of the Feminine Parts, and the Receptacle of the Womb, most exactly fitted for receiving

receiving the *Embryo*, and for cherishing and quickning it, will not persuade that what was made was made by Reason and Prudence? That the Woman should only differ from the Man in those Parts whereby Posterity was to be provided for and secur'd? As also, that the Frame of the Man should be different from that of the Woman, in those Parts only wherein the Power of Semination and Generation does reside? And herein certainly there is an illustrious Testimony of Providence afforded us; I mean in this necessary Diversity of the Parts. But yet this Testimony is stronger where we find an External Resemblance, and yet a Difference as to Use, and a Variety as to Operation. For so it is in the Paps, which are both in Men and Women; yet so that those of Women alone are capable to receive Milk, in order to the Infant's finding a proper Nutriment as soon as it is Born. Now therefore, if we see the Members dispos'd in Men with so great Skill, that while the Shape of all the other Parts is the same, those alone do admit of a Difference, wherein the several Uses require that Diversity; and while there is nothing in a Man that is superfluous or wanting, nor any Thing in a Woman that is too little or too much; who is there that does not evidently conclude from all these Observations, that all is the Effect of Reason, and of the Wisdom of the Creator?

The same Thing is confirm'd by the agreeable Diversity there is among other Animals, viii. 33.

mals, every one of which are suited to their proper Use and Service. This is also confirm'd from the Variety there is in Trees; the Diversity there is in Herbs; with the Difference of their Species and Juices: As it is also from the Changes of the Seasons of the Year, as distinguish'd into Four Parts, one succeeding another; from the regular Succession of Hours, Days, and Months in the annual Period; which Period never exceeds its appointed Limits one single Hour. Hence, Lastly, is it that the Age of the World it self is to be estimated at a certain Number of Years, without any Variation.

viii. 44.
But you will say, When was the World made? And why so lately? This you might as well have pretended, though it had been made sooner; for you might still have said, Why not sooner yet? For when you had gone backwards never so many Ages, you might always ask, Why not sooner still? But we are not now discoursing of this Matter, Why it was not made sooner than it was made; but whether it were made at all or not. For if it fully appears to have been made, 'twas certainly the Work of a Powerful and Supreme Artificer: Which when it is once settled, we must leave it to the Disposal and Judgment of the wise Artificer, when he thought fit to make it. Unless you will suppose, that all the Wisdom which fram'd this vast Structure of the World, and form'd all the distinct Sorts and Species of Beings, so as to dispose their Constitutions

not only to be agreeable in Point of Beauty, but withal most suitable and necessary for the Uses they were to be put to afterward, was only uncapable of this one Thing; I mean of chusing a proper Time for the rearing so magnificent a Building. Certainly he is not at a loss for sufficient Reasons, and evident Causes, why, and when, and how he would make the World; which were not surely to be reveal'd to Men, while they are scarce able to enquire after, and understand these Things that are before their Eyes, and are Testimonials of his Providence. For what is conceal'd in private, and is reposited within the wisest Understanding, as within a Royal Treasury, is disclos'd to none but to those who have learn'd from him with whom they are intrusted and reposited. 'Tis God therefore who made all Things, and was himself made by none. But for those that put the Name of *Nature*, for that of *God*; and so affirm, that all Things were made by *Nature*, they do not perceive the Mistake about that Appellation. For if they suppose this Nature to be Irrational, 'tis egregious Folly to imagine that a Creation where Reason is so visible, should proceed from a Maker who is destitute of it. But if this Nature be Reason, or the Word, whereby 'tis evident all Things were made, they chuse another Name to no Purpose; while they profess that he that created them is endued with Reason.

N.B. The following Testimonies, from the ancient Heathen Writers, are generally taken from the very Learned Dr. *Cudworth*'s *Intellectual System of the Universe*, and that nearly as he has translated them; where the Originals of them may also be consulted by the inquisitive Reader. A few others here are added out of the *Sibylline* Oracles, and from those Two most diligent and useful Naturalists, Mr. *Ray* and Mr. *Derham*; it being perfectly needless to make a new and larger Collection of my own out of the ancient Authors themselves, in so known and so endless a Matter as this is. At the Conclusion I have omitted most of the Christian Writers, as here of less Force, and as without Number; excepting a very few of the most eminent of our modern Philosophers; who were of the Laity also; and so on all Accounts truly unexceptionable Witnesses in this Case.

Cudworth p. 249.

Orpheus.] We will first sing a pleasant and delightful Song concerning the ancient *Chaos*; How Heaven, Earth and Seas were framed out of it: As also concerning that much-wise and sagacious Love, the oldest of all, and Self-Perfect, which actually produced all these things, separating one thing from another.

p. 300.

First of all, the Æther was made by God, and after the Æther a Chaos, a dark and dreadful Night, then covering all under the whole Æther.

—— *Orpheus* having declared also in his Explication, that there was a certain incomprehensible Being, which was the Highest and Oldest of all things, and the Maker of every thing, even of the Æther it self, and of all things under the Æther. But the Earth being then invisible, by reason of the Darkness, a Light breaking out through the Æther illuminated the whole Creation.

tion. This *Light* being said by him to be that Highest of all Beings, (beforementioned,) which is called also *Counsel*, and *Life*; these three Names in *Orpheus* (*Light*, *Counsel*, and *Life*,) declaring one and the same Force and Power of that GOD who is the Maker of all, and who produceth all out of nothing into Being, whether visible or invisible.

Wherefore, together with the Universe, were made within *Jove*, the Height of the Æthereal Heaven, the Breadth of the Earth and Sea, the great Ocean, the profound *Tartara*, the Rivers and Fountains, and all the other things, all the immortal Gods and Goddesses: Whatsoever hath been or shall be was at once contained in the Womb of *Jove*. p. 303.

The high thundring *Jove* is both the first and the last, *Jove* is both the Head and the Middle of all things: All things were made out of *Jove*. *Jove* is the Profundity of the Earth, and Starry Heaven; *Jove* is the Breath of all things; *Jove* is the Force of the untameable Fire; *Jove* is the Bottom of the Sea; *Jove* is the Sun, Moon, and Stars; *Jove* is both the Original and King of all things. There is One Power, and One God, and One great Ruler over all. *See page* 304, 305. p. 304.

Thales.] *Thales* said, that Water was the first Principle of all Corporeal things; but that GOD was that Mind which formed all things out of Water. p. 21.

Pythagoras.] *Pythagoras* thought, that GOD was a Mind passing through the whole Nature of Things; from whom our Souls were, as it were, cut off. p. 373.

Behold we see clearly, that *Pythagoras* held there was One GOD of the whole Universe, p. 377

the Principle and Cause of all things, the Illuminator, Animator and Quickner of the whole, and the Original of Motion; from whom all things were deriv'd, and brought out of Non-entity into Being.

p. 233.
Onatus.] It seemeth to me that there is not One GOD only, but that there is *One*, the greatest and highest God, that governeth the whole World, and that there are *Many* other Gods besides him, differing as to Power: That One GOD reigning over them all, who surmounts them all in Power, Greatness and Virtue. This is that GOD who contains and comprehends the whole World; but the other Gods are those, who, together with the Revolution of the Universe, orderly follow that first and intelligible GOD.

p. 396.
They who maintain that there is only one GOD, and not many Gods, are very much mistaken; as not considering aright, what the Dignity and Majesty of the Divine Transcendency chiefly consisteth in; namely in Ruling and Governing those which are like to it, and in excelling and surmounting others, and being superior to them. But all those other Gods which we contend for, are to that first and intelligible GOD as but the Dancers to the *Coryphæus* or *Choragus*, and as the inferior common Soldiers to the Captain or General; to whom it properly belongs to follow and comply with their Leader and Commander. The Work indeed is common, or the same to them both; to the Ruler and them that are Ruled; but they that are Ruled could not orderly conspire and agree together into one Work, were they destitute of a Leader; as the Singers and Dancers could not conspire together into one Harmony and Dance, were they destitute of a
Coryphæus;

Coryphæus; nor Soldiers make up an orderly Army, were they without a Captain or Commander.

Epicharmus.] Nothing is concealed from the Divinity: This well deserves your Knowledge. He is the Inspector of us. Nothing is impossible with GOD. p. 263.

Philolaus.] GOD is the Prince and Ruler of all, always one, stable, immoveable, like to himself, but unlike to every thing else. p. 393.

Archytas.] Whosoever is able to reduce all Kinds of things under one and the same Principle, this Man seems to me to have found out an excellent *Specula*, or high Station; from whence he may be able to take a large View and Prospect of GOD, and of all other things; and he shall clearly perceive that GOD is the Beginning, and End, and Middle of all things that are performed according to Justice and Right Reason. ibid.

Xenophanes.] There is one GOD, the greatest both among Gods and Men.—— He moveth the whole World without any Labour or Toil, meerly by Mind. *Theophrastus* affirms, that *Xenophanes* the *Colophonian, Parmenides* his Master, made one Principle of all things; he calling it *One and All,* and determining it to be neither finite nor infinite, (in a certain Sense.) and neither moving nor resting. Which *Theophrastus* also declares, that *Xenophanes* in this did not write as a Natural Philosopher, or Physiologer, but as a Metaphysician, or Theologer only. *Xenophanes* his *One and All,* being nothing else but GOD, whom he proved to be one solitary Being from hence, because GOD is the best and most powerful of all things; and there being many Degrees of Entity, there must needs be something Supreme to p. 377, 378.

Rule over all; which best and most powerful Being can be but one; he also did demonstrate it to be unmade, as likewise to be neither finite nor infinite, (in a certain Sense,) as he removed both Motion and Rest from GOD. Wherefore when he saith that GOD always remaineth or resteth the same, he understands not this of that Rest which is opposite to Motion, and which belongs to such things as may be moved, but of a certain other Rest which is both above that Motion and its Contrary.

p. 379. *Heraclitus.*] O you Unwise and Unlearned, teach us first what GOD is, that so you may be believ'd in accusing me of Impiety. Tell us where GOD is. Is he shut up within the Walls of Temples? Is this your Piety, to place GOD in the dark, or to make him a stony GOD? O you unskilful! know ye not that GOD is not made with Hands, and hath no *Basis* or *Fulcrum* to stand upon, nor can be inclosed within the Walls of any Temple? The whole World, variously adorn'd with Plants, Animals and Stars, being his Temple. —— Am I impious, O *Euthycles!* who alone know what GOD is? Is there no GOD without Altars? or are Stones the only Witnesses of him? No, His own Works give Testimony to Him, and principally the Sun; Night and Day bear witness of Him; the Earth bringing forth Fruits declares Him; The Circle of the Moon, that was made by Him, is an Heavenly Testimony of Him.

p. 505. *Heraclitus*'s Description of GOD is this, That most subtle and most swift Substance which permeates and passes through the whole Universe; by which all created things were made.

Zoroastres.]

Zoroastres.] GOD is the firſt, incorruptible, eternal, unmade, indiviſible, unlike to every thing elſe, the Head and Leader of all Good, one that cannot be bribed, the Beſt of the Good, the Wiſeſt of the Wiſe: He is alſo the Father of Equity and Juſtice, Self-taught, Perfect, and the only Inventor of what is naturally Holy.   p. 291.

Anaxagoras.] *Anaxagoras* affirmed, that there was, beſides Atoms, an ordering and diſpoſing Mind, that was the Cauſe of all things; ---which was the only ſimple, unmixed, and pure thing in the World.   p. 26.

He was the firſt, (that is, among the *Ionick* Philoſophers) who brought in Mind and GOD to the *Coſmopœia*; and did not derive all things from ſenſeleſs Bodies. Mind, the firſt Maker of the World; Mind, that which ſtill governs the ſame; the King and Supreme Monarch of Heaven and Earth.   p. 380.

Mind is mingled with nothing, but is alone by it ſelf, and ſeparate: For if it were not by it ſelf, diſtinct from Matter, but mingled therewith, it would then partake of all things; becauſe there is ſomething of all in every thing; which things mingled together with it would hinder it; ſo that it could not maſter or conquer any thing, as if alone by it ſelf. For Mind is the moſt ſubtile of all things, and the moſt pure, and has the Knowledge of all things, together with an abſolute Power over all.   p. 381.

Parmenides.] He deſcribes the Supreme Deity as the *one and all*, immutable; as one ſingle, ſolitary, and moſt ſimple Being; unmade, or ſelf-exiſtent, and neceſſarily exiſtent, incorporeal, and devoid of Magnitude, altogether immutable, or unchangeable, whoſe Duration therefore was very   p. 388.

ry different from that of ours, and not in a Way of Flux, or temporary Succession, but a *constant Eternity.*

p. 389, 390.
Parmenides, Melissus, and *Xenophanes.*] Perhaps, says *Simplicius,* it will not be improper for us to digress a little here, and to gratify the studious and inquisitive Reader, by shewing how those ancient Philosophers, tho' seeming to dissent in their Opinions concerning the Principles of the Universe, did notwithstanding harmoniously agree together. As first, of all, they who discoursed concerning the intelligible and first Principle of all, *Xenophanes, Parmenides,* and *Melissus*; of whom *Parmenides* called it *one, finite,* and *determined,* because as Unity must needs exist before Multitude, so that which is to all things the Cause of Measure, Bound, and Determination, ought rather to be describ'd by Measure and what is Finite, than by Infinity; as also that which is every way Perfect, and hath attained its own End, or rather is the End of all things, (as it was the Beginning,) must needs be of a determinate Nature: For that which is imperfect, and therefore indigent, hath not yet attain'd to its Term or Measure. But *Melissus,* though he considered the Immutability of the Deity likewise, yet attending to the inexhaustible Perfection of its Essence, the Unlimitedness and Unboundedness of its Power, declareth it to be Infinite, as well as Unbegotten or Unmade. Moreover *Xenophanes* looking upon the Deity as the Cause of all things; and above all things, placed it above Motion, and Rest, and all those Oppositions of inferior Beings; as *Plato* likewise doth in the first Hypothesis of his *Parmenides.* Whereas *Parmenides*

des and *Melissus* attending to its Stability, and constant Immutability, and its being perhaps above Energy and Power, praised it as immoveable.

Zeno Eleates.] *Zeno*, says *Aristotle*, by his one *Ens* which neither was moved, nor moveable, meaneth GOD. —— If GOD be the best of all things, then He must needs be One. ---- This is GOD, and the Power of GOD, to prevail, conquer and rule over all. Wherefore by how much any thing falls short of the Best, by so much does it fall short of being GOD. Now if there be supposed more such Beings, whereof some are better, some are worse; these could not be all Gods, because it is essential to GOD not to be transcended by any. But if they be conceiv'd to be so many equal Gods, then would it not be the Nature of GOD to be the Best: One Equal being neither better nor worse than another. Wherefore if there be a GOD, and this be the Nature of Him, then can there be but One. And indeed otherwise He could not be able to do whatsoever he would.

p. 390, 391.

Empedocles.] He is happy who hath his Mind richly fraught and stored with Treasures of Divine Knowledge; but he miserable, whose Mind is darkned, as to the Belief of a GOD.---- He denied GOD to be Corporeal;----- and affirmed that he is only an Holy and Ineffable Mind, that by swift Thoughts agitates the whole World.

p. 26.

Ecphantus and *Arcesilas.*] *Ecphantus* and *Arcesilas* held the corporeal World to consist of Atoms; but yet to be ordered and governed by a Divine Providence.

ibid.

Many of the oldest Philosophers.] It was a most ancient, and in a Manner universally received

p. 248.

ceived Tradition among the *Pagans*, that the *Cosmogonia*, or Generation of the World, took its first Beginning from a *Chaos*: This Tradition having been delivered down from *Orpheus*, and *Linus*, by *Hesiod*, and *Homer*, and others; acknowledged by *Epicharmus*, and embraced by *Thales*, *Anaxagoras*, *Plato*, and other Philosophers; and the Antiquity whereof is declared by *Euripides*.

p. 363, 364.

Euripides.] *Euripides*'s Prayer is, That GOD would infuse Light into the Souls of Men, whereby they might be enabled to know what is the Root from whence all their Evils spring, and by what Means they may avoid them. And elsewhere,

Thou self-sprung Being, that dost all enfold,
And in thine Arms Heav'ns whirling Fabrick hold,
Who art encircled with resplendent Light,
And yet ly'st mantled o're in shady Night;
About whom the exultant Starry Fires
Dance nimbly round in everlasting Gyres.

p. 353.

Sophocles.] There is in Truth one only GOD, who made Heaven and Earth, the Sea, Air, and Winds, &c.

p. 399.

Socrates.] I am now convinced, says *Aristodemus* to *Socrates*, from what you say, that the Things of this World were the Workmanship of some wise Artificer, who also was a Lover of Animals: ——— Do you think that you only have Wisdom in your self, and that there is none else in the World without you? ——— Is Mind and Understanding therefore the only thing which you fancy you have some way or other luckily got, and snatched unto your self, whilst there

there is no such thing any where in the World without you? All those infinite things therein being thus orderly disposed by Chance?----- Neither do you see your own Soul which rules over your Body: So that you might for the same Reason conclude your self to do nothing by Mind and Understanding neither, but all by Chance, as well as that all things in the World are done by Chance. ------ How much the more magnificent and illustrious that Being is which takes care of you, so much the more in all Reason ought it to be honoured by you. ------ Consider, Friend, I pray you, if that Mind which is in your Body does order and dispose it every way as it pleases, why should not that Wisdom which is in the Universe be able to order all things therein also as seemeth best to it? And if your Eye can discern things several Miles distant from it, why should it be thought impossible for the Eye of GOD to behold all things at once? Lastly, if your Soul can mind things both here, and in *Egypt*, and in *Sicily*, why may not the great Mind or Wisdom of GOD be able to take care of all things, in all Places? -----GOD is such and so great a Being, as that he can at once see all things, hear all things, and be present every where, and take care of all Affairs. ----The other Gods giving us good things, do it without visibly appearing to us; and that GOD who framed and containeth the whole World, in which are all good and excellent things, and who continually supplieth us with them, He, though He be seen to do the greatest things of all, yet notwithstanding is Himself invisible and unseen: Which ought the less to be wondered at by us, because the Sun, which is manifest to all, yet

will

will not suffer himself to be exactly and distinctly view'd; but if any one boldly and impudently gaze upon him, will deprive him of his Sight. As also because the Soul of Man, which most of all things in him partaketh of the Deity, tho' it be that which manifestly rules in us, yet it is never seen. Which Particulars he that considers, ought not to despise invisible things, but to honour the Supreme Deity, taking Notice of his Power from his Effects.

p. 63. *Plato.*] Whosoever had but the least of Seriousness and Sobriety in them, whensoever they took in hand any Enterprize, whether great or small, they would always invoke the Deity for Assistance and Direction.

p. 155. Those things which are said to be done by Nature, are indeed done by Divine Power.

p. 402. When I begin my Epistles with *GOD*, then may you conclude I write seriously; but not so when I begin with Gods.

p. 404. *Plato* calls the Supreme GOD, *The GOD*: The Architect or Artificer of the World; the Maker and Father of this Universe; whom it is hard to find out, but impossible to declare to the vulgar: The GOD over all; The Creator of Nature: The sole Principle of the Universe: The Cause of all things: Mind, the King of all things: That Sovereign Mind which orders all things, and passes through all things: The Governor of the whole: That which always is, and was never made: The First GOD: The greatest GOD, and the greatest of the Gods: He that governeth or produceth the Sun: He that makes the Earth, and Heaven, and the Gods, and doth all things both in Heaven, and Hell, and under the Earth.

Aristotle.]

Aristotle.] *Aristotle* plainly affirms, that all the Philosophers before himself did assert the World to have been made, or have had a Beginning. p. 118.

There is more of Design or final Cause, and of wise Contrivance, in the Works of Nature, than in those of human Art. p. 165.

It is more probable that the whole World was at first made by Art, (if at least it were made,) and that it is still preserved by the same, than that mortal Animals should be so. For there is much more of Order and determinate Regularities in the heavenly Bodies, than in our selves; but more of Fortuitousness and Want of Regularity among these mortal things. Notwithstanding which, some there are who, tho' they cannot but acknowledge that the Bodies of Animals were all framed by an artificial Nature, yet they will needs contend that the System of the Heavens sprung meerly from Fortune and Chance; altho' there be not the least Appearance of Chance or Incogitancy in it. p. 168.

If there be any such Substance as this that is separate (from Matter, or Incorporeal) and immoveable (as we shall afterwards endeavour to shew that there is;) then the Divinity ought to be placed here; and this must be acknowledged to be the first and most proper Principle of all. p. 386.

All Men have an Opinion or Persuasion that there are Gods. And they who think so, as well *Barbarians*, as *Greeks*, attribute the highest Place to that which is Divine; as supposing the immortal Heavens to be most accommodate to immortal Gods. p. 409.

Unless there were something else in the World besides what is *sensible*, there could be neither Beginning nor Order; but one thing would be the Principle p. 413.

Principle of another infinitely, or without End. ----It is not at all likely that either Fire, or Earth, or any such Body should be the Cause of that Fitness or Proportion that is in the World. Nor can so noble an Effect as this be reasonably imputed to Chance or Fortune.

p. 475, 476.
Aratus.] Let us begin with *Jove*: Him of whom we Men are never silent; and of whom all things are full. He penetrating and pervading all, and being every where; and whose Beneficence we constantly make use of and enjoy; for we also are his Off-spring; who as a kind and benign Father sheweth lucky Signs to Men: For he also fixeth the Signs in Heaven, distinguishing Constellations, and appointing Stars to rise and set at several Times of the Year. Therefore is he always propitiated, and appeased both first and last. Hail, O Father! the great Wonder of the World, and the Interest of Mankind.

p. 432, 33, 434.
Cleanthes.] *Cleanthes*'s Prayer to the Supreme GOD. ' Hail *Jove*! most glorious of the im-
' mortal Beings, who hast many Names, and art
' ever Omnipotent, the Author of Nature, go-
' verning all things by Law. For all Mortals are
' allowed to address to thee. For we are thy Off-
' spring; tho' a meer Imitation or Eccho of Thee,
' even all of us who live and creep upon the
' Earth. Wherefore I will sing an Hymn to
' Thee, and always Praise Thy Power. All this
' round World that circles about the Earth obeys
' Thee, whithersoever Thou guidest it, and vo-
' luntarily submits to Thy Government. Such
' two-edged, fiery, and ever-living, Thunder hast
' Thou, as Thy Instrument, under Thy victori-
' ous Hands; for all Nature trembles under Thy
' Stroke; by the same dost Thou rule that com-

'mon Reason [or Word] which penetrates
'through all Things. So great and Supreme a
'King art Thou always; nor is there, O Spirit,
'any Work done upon Earth without Thee,
'nor in the æthereal and divine Heaven, nor in
'the Sea, but what the Wicked do by their own
'Folly. What is disordered Thou reducest in-
'to Order, and what is inimical Thou rendrest
'friendly to Thee. In such a Manner dost
'Thou adjust the confused State of things,
'good and bad, that there arises a rational Sy-
'stem of Beings, perpetually going on, which
'all the Wicked avoid, and will not acquiesce
'in: Miserable as they are, who still, tho' de-
'sirous of the Enjoyment of Happiness, have no
'Regard to the common Law of GOD, nor will
'hearken thereto; which if they would submit
'to they might enjoy a sound Mind, and a hap-
'py Life: But they indeed do without Goodness
'bend their Inclinations to several things; some
'affect the troublesome Toil of Ambition; others
'turn themselves to Covetousness, without any
'Regard to Decency; others give themselves to
'Rest, and the Pleasures of the Body. But do
'Thou, O *Jove*, the Giver of all things, who
'inhabitest in the dark Clouds, and governest
'the Thunder, Deliver Men from their foolish
'and unhappy Inclinations, and drive such Pas-
'sions away from their Souls. Grant them Skill
'to understand this Thy Determination, accord-
'ing to which thou with Justice governest all
'things; that so we, finding our selves honoured
'of Thee, may pay back the Tribute of our
'Honour to Thee, by celebrating thy Works in
'our Hymns perpetually, as it becomes a Mor-
'tal Being to do: For there is not any nobler
                                            ' thing

'thing which either mortal Men, or the Gods
'themselves can be employ'd in, than to cele-
'brate righteously by Hymns the common Law
'of the entire System.

p. 255. *Cicero.*] The Entire Nature or the Universe is governed by the Force, Reason, Power, Mind, and Divinity of the Immortal Gods.

p. 256. The Minds of Citizens ought to be first of all embued with a firm Persuasion that the Gods are the Lords and Moderators of all things, and that the Conduct and Management of the whole World is directed and over-ruled by their Judgment and Divine Power; that they deserve the best of Mankind; that they behold and consider what every Man is, what he doth, and takes upon himself; with what Mind, Piety, and Sincerity he observes the Duties of Religion; and, lastly, that these Gods have a very different Regard to the pious and the impious.

p. 434. That there is some most excellent and eternal Nature, which is to be admired and honoured by Mankind, the Beauty of the World, and the Order of the Heavenly Bodies compel us to confess.

ibid. Who is so mad or stupid, as when he looks up to Heaven is not presently convinced there are Gods? or can perswade himself, that those things which are made with so much Mind and Wisdom, as that no human Skill is able to reach and comprehend the Artifice and Contrivance of them, did all happen by Chance?

p. 435. I say that the World, and all its Parts, were at first constituted by the Providence of the Gods.

p. 436. We must needs acknowledge that the Benefits of this Life, the Light which we enjoy, and the Spirit which we breathe, are imparted to us from GOD.

There is, there is certainly such a Divine Force in the World. Neither is it reasonable to think that in these gross and frail Bodies of ours, there should be something which hath Life, Sense, and Understanding: and yet no such thing in the whole Universe. Unless Men will therefore conclude that there is none, because they see it not: As if we could see our own Mind, (whereby we order and dispose all things, and whereby we reason and speak thus,) and perceive what kind of thing it is, and where it is lodged. *ibid.*

Neither can GOD himself be understood by us otherwise than as a certain distinct and free Mind, separate from all mortal Concretion, which both perceives and moves all things. *ibid.*

When we behold these and other wonderful Works of Nature, can we at all doubt but that there presideth over them either One Maker of all, if they had a Beginning, as *Plato* conceiveth; or else, if they always were, as *Aristotle* supposeth, One Moderator and Governor? *ibid.*

Without Government neither any House, nor City, nor Nation, nor Mankind in general, nor the entire Nature of things, nor the World it self could subsist. For this also obeyeth GOD; and the Seas and Earth are subject to Him, and the Life of Man is disposed of by the Commands of the Supreme Law. *ibid.*

Whosoever thinketh that the admirable Order, and incredible Constancy of the Heavenly Bodies, and their Motions, whereupon the Preservation and Welfare of all things doth depend, is not govern'd by Mind and Understanding, he himself is to be accounted void thereof. -------- Shall we, when we see an artificial Engine, as a Sphere, or Dial, or the like, at first Sight acknowledge that *Idem ap. Ray, of the Creation, p. 67, 68.*

P

that it is a Work of Reason and Art? And can we, when we see the Force of the Heavens, mov'd and carried about with admirable Celerity, most constantly finishing its annual Revolutions, to the eminent Welfare and Preservation of all things, doubt at all that these things are perform'd not only by Reason, but by a certain excellent and divine Reason?

Idem ap. Ray, of the Creation, p. 221, 222.

We might add many Reasons to this of the Providence, and Diligence, and Sagaciousness of Nature on our Account; whereby we may be satisfied how great and eminent Blessings are bestowed on Men by GOD; who at first raised them from the Ground, and set them in an erect and upright Posture, that by viewing the Heavens they might attain to the Knowledge of the Gods. For Men are elevated from the Earth, not like Inhabitants, but like Spectators of Heavenly things; the Consideration of which belongs to no other Sort of Animals.

Idem ap. Derham; *Astrotheol.* p. 4, 5.

What can be so plain and clear as when we behold the Heavens, and view the Heavenly Bodies, that we should conclude there is some Deity of a most excellent Mind by which these things are govern'd? ---- a present and an Almighty God; which he that doubts of, I do not understand why he should not as well doubt whether there be a Sun or no. ---- Time wears out the Figments of Opinions, but confirms the Judgments of Nature; for which Reason both among our selves, and in other Nations, the Veneration of the Gods, and the Sacredness of Religion, augment and improve every Day more and more.

ibid. *Idem,* p. 60, 61.

If thou should'st see a large and fair House, thou could'st not be brought to imagine that House was built by the Mice and Weezles; altho' thou should'st shouldst not see the Master thereof. So would'st thou not think thy self very plainly to play the Fool, if thou should'st imagine so orderly a Frame of the World, so great a Variety and Beauty of Heavenly Things, so prodigious a Quantity and Magnitude of Sea and Land, to be thy House, thy Workmanship, and not that of the Immortal Gods?

The fourth Cause, and that even the chief, is the Equality of the Motion and Revolution of the Heavens; the Distinction, Unity, Beauty, and Order of the Sun, Moon, and all the Stars; the bare View alone of which Things is sufficient to demonstrate them to be no Works of Chance. As if any one should come into an House, the *Gymnasium*, or *Forum*, when he should see the Order, Manner, and Management of every Thing, he could never judge these Things to be done without an Efficient; but must imagine there was some Being presiding over them, and whose Orders they obeyed. Much more in so great Motions, such Vicissitudes, and the Orders of so many and great Things: ----- A Man cannot but conclude that such great Acts of Nature are governed by some Mind. *Idem, ibid. p. 105, 106.*

So the Philosophers ought to have done, if haply they had any Doubts at the first View of the World; afterwards, when they should behold its Determinate and Equal Motions, and all Things managed by established Orders, and with immutable Constancy, they ought then to understand, that there is not only some Inhabitant in this Heavenly, this Divine House; but also some Ruler and Moderator, and in a manner Architect of so great a Work, so noble a Performance. *Idem, ibid. p. 72.*

Idem, ibid. p. 212, 213.

It was the Opinion of *Aristotle*, that if there were such a Sort of People that had always lived under the Earth, in good and splendid Habitations, adorned with Imagery and Pictures, and furnished with all Things that those accounted happy abound with; and supposing that these People had never at any Time gone out upon the Earth, but only by Report had heard there was such a Thing as the Deity, and a Power of the Gods; and that at a certain Time afterwards the Earth should open, and this People get out from their hidden Mansions into the Places we inhabit, when on the sudden they should see the Earth, the Seas, and the Heavens; perceive the Magnitude of the Clouds, and the Force of the Winds; behold the Sun, and its Grandeur, and Beauty, and know its Power in making the Day, by diffusing its Light through the whole Heavens; and when the Night had overspread the Earth with Darkness, they should discern the whole Heavens bespread and adorned with Stars, and see the Variety of the Moon's Phases, in her Increase and Decrease, together with her Risings and Settings, and the stated and immutable Courses of all these throughout all Eternity; this People, when they should see all these Things, would infallibly imagine that there are Gods, and that those grand Works were the Works of the Gods.

Cudw. p. 439.

Varro.] These alone seem to *Varro* to have understood what GOD is, who believed him to be a Soul, governing the whole World, by Motion and Reason.

Proæm. pag. 3, 5.

The Sybilline Oracles.] O Mortal, Carnal, and Vile Men! How soon are you puft up? not considering that you must die. You don't tremble at, at, and fear the Supreme GOD who governs you, who knows, sees, and observes all Things; who is the Creator that preserves all Things; who sent his pleasant Spirit into all Things; and made him the Governor of all Mankind. There is One GOD, who alone Reigns: He is very Great, Unbegotten, Omnipotent, Invisible. He alone sees all Things, but cannot be seen by any Mortal; for what Flesh can behold the Celestial, True, and Immortal GOD with his Eyes, who lives in Heaven? since Men who are born Mortals, of Bones, Flesh, and Veins, cannot stedfastly behold the shining Beams of the Sun. Worship Him who is the Only Governor of the World; who alone is from Everlasting to Everlasting: He exists from himself, is Unbegotten; He Governs all Things at all Times, and He hath ordained a Judgment for all Men in one common Day.----Behold, He is manifest to all, and is no Deceiver. Come therefore, and do not pursue this dark and tempestuous way of [Idolatry and Polytheism] any longer. Behold the pleasant Light of the Sun shines gloriously. Know, and wisely consider it, there is One GOD, who gives Rain, and Winds; He causes Earthquakes, Thunders, Famines, Plagues, Snow, Ice, and such grievous Calamities. But why do I reckon up every Particular? He commands in Heaven, and governs in the Earth, and really exists. ----- p. 5.

There is One only Supreme GOD, who hath created Heaven, the Sun, Moon, and Stars, and the fruitful Earth, and the swelling Waves of the Sea, the Mountains full of Woods, and the eternal Streams of the Fountains. He produceth an innumerable Quantity of Fish in the Waters; and He nourisheth the creeping Creatures with p. 5, 7.

with a cool Diet: And He gives to the swift Birds of various Kinds, harsh and pleasant Notes, and to cut the Air with their noisy Wings; and He hath put the wild Beasts in the Hills, covered with Wood; and hath subjected all Beasts to Mankind: But hath made Man His peculiar Workmanship, the Governor of all Things; and hath subjected to him many various Creatures which he cannot comprehend: For what mortal Man can know all Things? But He only knows them who made them in the Beginning; who is the incorruptible and eternal Creator, living in Heaven; who gives to all Good Men a very great Reward; but is angry with the Unjust, and Wicked, and punishes them by Wars, Plagues, and extraordinary Calamities.

Cudw. p. 251.

Diodorus Siculus, of the old *Chaldeans*.] The *Chaldeans* affirm the Nature of the World to be eternal; and that it was neither generated from any Beginning, nor will ever admit Corruption. They believe also, that the Order and Disposition of the World, is by a certain Divine Providence; and that every one of those Things which come to pass in the Heavens, happens not by Chance, but by a certain determinate and firmly ratified Judgment of the Gods.

Ovid.] Of the Creation, see the Beginning of his *Metamorphoses* at large. Other Passages out of *Plautus*, *Virgil*, *Horace*, and the rest of the Poets are common, but here omitted for the Sake of Brevity, and to leave Room for other Testimonies.

p. 247.

Strabo.] *Strabo* affirms that the World was the joint Work of Nature and Providence; ----- which Providence, having a manifold Fecundity in it, and delighting in Variety of Works, it
designed

of RELIGION. 215

designed principally to make Animals, as the most excellent Things; and among them chiefly those Two noblest kinds of Animals, Gods, and Men; for whose Sakes the other Things were made; and then assigned Heaven to the Gods, and Earth to Men, the two extreme Parts of the World, for their respective Habitations.

Strabo testifies of the ancient *Indian Brachmans*, that in many Things they philosophiz'd after the *Grecian* Manner; as when they affirm, that the World had a Beginning, and that it would be corrupted, and that the Maker and Governor thereof pervades the whole of it. *p. 504.*

Seneca.] GOD, when He laid the Foundation of this most beautiful Fabrick, and began to erect that Structure, than which Nature knows nothing greater or more excellent; to the End that all Things might be carried on under their respective Governors orderly, though he intended himself through the whole, as to preside in chief over all; yet did He *generate Gods* also, as subordinate Ministers of His Kingdom under Him. *p. 247.*

Seneca calls GOD, the Framer and Former of the Universe; The Governor, Disposer, and Keeper thereof; Him upon whom all Things depend: The Mind and Spirit of the World: The Artificer and Lord of this whole mundane Fabrick: To whom every Name belongs: From whom all Things spring: By whose Spirit we live: Who is in all His Parts, and sustaineth Himself by His own Force: By whose Counsel the World is provided for, and carried on in its Course, constantly and uninterruptedly: By whose Decree all Things are done: The Divine Spirit that is diffus'd through all Things, both great and small, in an equal Degree: The GOD whose *p. 440.*

P 4

whose Power extends to all Things: The Greatest and most Powerful GOD, who doth Himself support and uphold all Things: Who is present every where to all Things: The GOD of Heaven, and of all the Gods; upon whom are suspended all those other Divine Powers which we singly worship and adore.

Mr. *Derham*'s Astrotheol. p. 217. ex Epist. 117.

Seneca instanceth in two Things that have the Consent of Mankind for them; the Immortality of the Soul; and the Existence of the Deity: Which, saith he, among other Arguments, we collect from the innate Opinion which all Men have of the Gods. For there is no Nation in the World so void of Law and Morality, as not to believe but there are some Gods.----- They lie, that say they believe there is no GOD: For atho' by Day they may affirm so to thee, yet by Night they are to themselves conscious of the contrary.

Quare bonis Viris, &c. C 1. ib. p. 217, 218.

Seneca takes it for granted, that there is such a Thing as a Divine Power and Providence, governing the World; and he saith, It was needless for him to shew that so great a Work [as the World] could not stand without some Ruler; that so regular Motions of the Stars could not be the Effects of a fortuitous Force; and that the Impulses of Chance must be oftentimes disturb'd and justle; that this undisturbed Velocity, which bears the Weight of so many Things in the Earth, and Seas, with so great a Number of Heavenly Lights, both very illustrious and also shining, according to a manifest Regularity, must needs proceed by the Direction of some Eternal Law: That this can never be the Order of straggling Matter; neither is it possible for Things fortuitously and rashly combin'd to depend upon, and manifest so much Art.   *Quintilian.*]

Quintilian.] GOD is a Spirit, mingled with, and diffus'd through all the Parts of the World. p. 440, 504.

Plutarch.] It is better for us to follow *Plato*, and loudly to declare that the World was made by GOD. For as the World is the best of Works, so is GOD the best of all Causes. Nevertheless, the Substance or Matter out of which the World was made, was not it self made, but always ready at hand, and subject to the Artificer, to be ordered and disposed by Him. For the making of the World was not the Production of it out of nothing, but out of an antecedent, bad and disorderly State; like the making of an House, Garment, or Statue. p. 197.

GOD seems to excel in these three Things; Incorruptibility, Power, and Virtue: Of all which the most divine and venerable is Virtue. For *Vacuum*, and the senseless Elements have Incorruptibility: Earthquakes and Thunder, Blustring Winds, and over-flowing Torrents, much of Power and Force: Wherefore the Vulgar being affected three Manner of ways towards the Deity, so as to Admire its Happiness; to Fear it; and to Honour it; they Esteem the Deity Happy, for its Incorruptibility: They Fear it, and stand in Awe of it for its Power; but they Worship it, that is, Love and Honour it, for its Justice. p. 203.

Whereas there are two Causes of all Generation, (the Divine and the Natural,) the most Ancient Theologers and Poets attended only to the more excellent of these two, (the Divine Cause;) resolving all Things into GOD, and pronouncing this of them universally, That Jove *was both the Beginning, and Middle, and that all Things were out of* Jove; [as the *Orphick* Verse has it.] p. 305.

Insomuch

Insomuch that these had no regard at all to the other natural and necessary Causes of Things. But on the contrary, their Juniors, who were called Naturalists, straying from this most excellent and divine Principle, placed all in Bodies, their Affections, Collisions, Mutations, and Mixtures together.

p. 423, 424.

Neither is it at all considerable what the Stoicks here object against a Plurality of Worlds; they demanding how there could be but one Fate, and one Providence, and one *Jove*, (or independent Deity,) were there many Worlds? For what Necessity is there that there must be more *Joves* than one, if there were more Worlds? And why might not that One and the same GOD of the Universe, call'd by us the Lord and Father of all, be the first Prince, and Highest Governor in all those Worlds? Or what hinders but that a Multitude of Worlds might be all subject to the Fate and Providence of one *Jove*, or Supreme GOD? Himself inspecting and ordering them every one, and imparting Principles and Spermatick Influences to them, according to which all Things in them might be governed and disposed. For can many distinct Persons in an Army, or *Chorus*, be reduc'd into one Body or Polity, and could not ten, or fifty, or a hundred Worlds in the Universe be all govern'd by One Reason, and be ordered together in reference to One Principle. [*See Page* 459.

Plut. de placit. Phil. l. 6.

Men began to acknowledge a GOD, when they saw the Stars maintain so great an Harmony, and the Days and Nights, both in Summer, and Winter, to observe their stated Risings and Settings.

Dion

Dion Chryſoſtom.] The whole World is under a Kingly Power, or Monarchy. ------ The Supreme GOD is the common King of Gods and Men, their Governor and Father: The GOD that rules over all: The Firſt and Greateſt GOD: The chief Preſident over all Things; who orders and guides the whole Heaven and World, as a wiſe Pilot doth a Ship: The Ruler of the whole Heaven, and Lord of the whole Subſtance of Things. ------ Concerning the Nature of the Gods in general, but eſpecially of that Supreme Ruler over all, there is an Opinion in all Human Kind, as well *Barbarians* as *Greeks*, that is naturally implanted in them, as rational Beings, and not deriv'd from any Mortal Teacher. {Cudw. p. 443, 444.}

Galen.] Should I any longer inſiſt upon ſuch Brutiſh Perſons as thoſe, the Wiſe and Sober might juſtly condemn me, as defiling this Holy Oration, which I compoſe as a true Hymn to the Praiſe of Him that Made us; I conceiving true Piety and Religion towards GOD to conſiſt in this, not that I ſhould ſacrifice many *Hecatombs*, or burn much Incenſe to Him; but that I ſhould my ſelf firſt acknowledge, and then declare to others, how great His Wiſdom is, how great His Power, and how great His Goodneſs. For that He would adorn the whole World after this Manner, envying to nothing that Good which it was capable of, I conclude to be a Demonſtration of moſt abſolute Goodneſs: And thus let Him be Praiſed by us as Good. And that He was able to find out how all Things might be adorn'd after the beſt Manner, is a Sign of the greateſt Wiſdom in Him. {p. 444.}

And

And, lastly, to be able to effect and bring to pass all those Things which he had thus decreed, argues an insuperable Power. [*See much more to this Purpose, in his admirable Book, De Usu Partium.*]

p. 445.

Maximus Tyrius.] I will now more plainly declare my Sense by this Similitude: Imagine in your Mind a great and powerful Kingdom or Principality, in which all the rest freely and with one Consent conspire to direct their Actions agreeable to the Will and Command of One Supreme King, the Oldest and the Best: And then suppose the Bounds and Limits of this Empire not to be the River *Halys*, nor the *Hellespont*, nor the Lake of *Mæotis*, nor the Shores of the Ocean; but Heaven above, and the Earth beneath. Here then let that great King sit Immoveable, prescribing Laws to all His Subjects; in which consists their Safety and Security: The Consorts of his Empire being many, both visible and invisible Gods: Some of which, that are nearest to him, and immediately attending on Him, are in the highest Royal Dignity, feasting, as it were, at the same Table with Him. Others again are their Ministers and Attendants; and a third Sort Inferior to them both. And thus you see how the Order and Chain of this Government descends down by Steps and Degrees from the Supreme GOD, to the Earth and Men.

p. 517.

The End of your Journey is not the Heaven, nor those shining Bodies in the Heaven; for tho' those be beautiful, and divine, and the genuine Off-spring of the Supreme Deity, ty, framed after the best Manner; yet ought all these to be transcended by you, and your Heart lifted up far above the starry Heavens, [to the Father and Maker of all Things.]

Macrobius.] The whole World is well cal- p. 538, led *the Temple of GOD*, in way of Oppo- 539. sition to those who think GOD to be nothing else but the Heaven it self, and those Celestial Things which we see. Wherefore *Cicero*, that he might shew the Omnipotence of the First and Supreme GOD to be such as could scarcely be understood, but not at all perceiv'd by Sense; He calleth whatsoever falleth under Human Sight, *His Temple*; that so he that worshippeth these Things, as the Temple of GOD, might in the mean time remember, that the chief Worship is due to the Maker and Creator of them. As also that himself ought to live in the World like His Priest; holily and religiously.

Jamblicus, of the old *Ægyptian* Theology.] p. 335, They assert that GOD, who is the Cause of 336. Generation, and of entire Nature, and of all the Powers in the Elements themselves, is separate, exempt, elevated above, and expanded over all the Powers and Elements in the World. For being above the World, and transcending the same, Immaterial, and Incorporeal, Above Nature, Unmade, Indivisible, manifested wholly from Himself, and in Himself, He rules over all things; and in himself containeth all Things; and because he virtually comprehends all Things, therefore does he impart and display the same form Himself.

Minutius

Translated by Mr. Reeves.

Minutius Felix, in part from the old Philosophers.] §. XVII. Nor do I deny, what *Cæcilius* has taken so much pains to prove, that Man must learn to know himself, and diligently examine his Nature, his Original, and the End of his Being; whether he was only a mere Concretion of the Elements, and thus admirably adjusted by blind Atoms; or made, and fashioned, and animated by God. But this we cannot apprehend without studying the World, and its Maker; for these Things are so closely connected and chain'd together, that you must diligently examine the Nature of God, before you can understand that of Man; nor can you ever be a good Citizen of the World, before this common City of us All, the World and You, are well acquainted; and certainly, since in this chiefly it is that we differ from Beasts, that whereas they are prone to Earth, and bent downward by Nature, and fram'd to look no farther than the good of their Bellies; yet Man is made erect and upright, and by that Make form'd for the Contemplation of Heaven, and has Language and Reason to conduct him to the Knowledge and Imitation of God; for a Creature so constituted to be ignorant of his Maker, to wink as hard as he can, that he may not see that Glorious Being that is thrusting in at his Eyes, and knocking for Admission at all his Senses, is the most inexcusable Ignorance imaginable. For 'tis most abominable Sacrilege to be Poring upon Earth, for that which you are only to find in Heaven. For which reason I can hardly think that such Men have the use of their Soul or Senses, no-not of their very Eyes, who cannot see this glorious Machine of the Universe to be the Work of Divine

vine Wifdom, but dream that 'twas jumbled together by a fortuitous Concourfe of Atoms. For what is fo clear and undeniable, when you lift up your Eyes to Heaven, and when you look down upon all about you, than that there is a Deity of moft excellent Underftanding, that infpires, moves, fupports, and governs all Nature? Confider the vaft Expanfe of Heaven, and the Rapidity of its Motion, either when it is ftudded with Stars by Night, or enlighten'd with the Sun by Day; then fhall you fee that Almighty Hand which poifes them in their Orbs, and balances them in their Movement. Behold how the Sun girds up and regulates the Year by its Annual Circuit, and how the Moon meafures round a Month by its Increafe, Decay, and total Difappearance. What need I mention the conftant Viciffitudes of Light and Darknefs, for the alternate Reparation of Reft and Labour? I muft leave it to Aftrologers to tell you more at large the Ufes of the Stars, either how they direft the Pilot in Navigation, or the Husbandman in his Seafons of Plowing and Reaping; every one of which Celeftial Bodies, as they requir'd Almighty Power and Wifdom at firft to create and range them in their Stations, fo do they require the moft confummate Wifdom and Sagacity to comprehend them now they are created. Moreover, does not the ftanding Variety of Seafons marching in goodly Order teftify the Divine Author? The Spring with her Flowers, the Summer with her Harvefts, and the Ripening Autumn with grateful Fruits, and the moift and unctuous Winter, are all equally neceffary; which Order had certainly been difturbed before now, had it not been fix'd by the

wifeft

wisest Power. What an Argument of a Providence is it, thus to interpose and moderate the Extremes of Winter and Summer, with the Allays of Spring and Autumn, that we pass the Year about with Security and Comfort, between the excesses of Parching Heat and Cold? Observe the Sea, and you'll find it bounded with a Shore, a Law it cannot transgress; look into the vegetable World, and see how all the Trees draw their Life from the Bowels of the Earth; view the Ocean in constant Ebb and Flow, and the Fountains running in full Veins, and the Rivers perpetually gliding in their wonted Channels. What need I spend more Words to shew, how providentially this Spot of Earth is canton'd out into Hills, and Dales, and Plains? What need I speak of the various Artillery for the Defence of every Animal? Some arm'd with Horns, or hedg'd about with Teeth, or fortify'd with Hoofs and Claws, or spear'd with Stings; and others either swift of Foot, or Wing? But above all, the beautiful Structure of Man most plainly speaks a God; Man of Stature straight, and Visage erect, with Eyes at top like Centries, watching over the other Senses within the Tower.

XVIII. But I shou'd never come to an end, was I to travel through Particulars; there is not any one Part in Man, but is either necessary or ornamental. And what is still more miraculous, is to find a general Resemblance in all, and distinguishing Features in each; so that the whole Species is alike, and yet not one Individual without some discriminating Character. What think you of the manner of our Birth, and the Instinct of Generation? Who but God cou'd cou'd turn the Course of Nature against such a Time, to fill the Breasts with Milk for the ripening *Embryo*, and suckle the tender Infant with that Plenty of lacteal Dew? Nor does God provide only for Universals, but takes care also of Particulars. *Britain* is made amends with the warm Vapours of the circumambient Sea for its deficiency of Sun. *Nile* serves *Egypt* for Rain. *Euphrates* cultivates *Mesopotamia*; and *Indus* is said both to water and sow the *East with the Seeds it discharges into it*. Shou'd you chance to go into a House, and see all the Rooms exquisitely furnish'd, and kept in great order, you wou'd make no dispute but such a House was under the Care and Inspection of a Master, and that he himself was preferable to all the Furniture. Thus in this Palace of the World, when you cast your Eyes upon Heaven and Earth, and behold the admirable Order and Oeconomy of Things, you have as little reason to question whether there is a Lord of the Universe, and that he himself is more glorious than the Stars, and more to be admir'd than the Works of his own Hands. But perhaps you may have no Scruples concerning a Providence, but only whether the Heavenly Government is lodg'd in One, or a Plurality of Deities. And this is easily decided, if you'll give your self but the Trouble to look abroad into the Kingdoms of the World, from which you may collect the Regimen or Form above. For when did you ever know any Copartnership in a Kingdom commence with Integrity, or conclude without Blood? Not to mention the Grandees of *Persia*, who consulted the Neighing of a Horse in the Election of Kings, nor to revive the old Story of the *Theban Pair* dead and gone; the fatal Diffention of the *Roman Brothers* for a Kingdom of Shepherds, and Shepherds Sheds, is famous all the World over. The Wars of *Father* and *Son-in-Law*, *Cæsar* and *Pompey*, shook the Earth; and all the *Roman* Empire was not big enough to hold Two Men. See Examples of another kind; the Bees have but one King, and the Flocks and Herds but one Leader; and can you imagine Two Supremes in Heaven, and that Almighty Power is divifible? Since 'tis manifeft, that God, the Univerfal Parent, has neither Beginning nor End; but gave Beginning to All, and Eternity to himfelf; who before the World was, was a World to himfelf; who commands all Things by his *Word*, and difpenfes them by his Wifdom, and confummates them by his Power. This God is Invifible, becaufe of his Brightnefs inacceffible; and not tangible, becaufe Incorporeal; and Incomprehenfible, becaufe too great for our Capacity; Infinite, Immenfe, and this Immenfity intelligible by himfelf only. Our Intellect is too narrow to contain him, and therefore we never conceive fo worthily of him, as when we conceive him Unconceivable. Shall I fpeak my Senfe of this Matter? Whoever imagines that he knows the Divine Majefty, leffens it; and whoever does not leffen it, can never pretend to know it. Enquire not his Name, for God is his Name, and there only we ufe Names, where many Individuals are to be diftinguifh'd by their proper Appellations; but to God, who is but One, the Name of God is all in all; for if I call him Father, you forthwith conceive of him, as an Earthly Parent; if

King

King or Lord, your Fancy cloths him with such Ideas as those Words stand for with Men. Take but away this human Covering of Words, and you'll see the Divine Nature the better. Moreover, have I not all the World on my Side in the Acknowledgment of this One God? I hear the People when they lift up their Hands to Heaven, say nothing else, but *The God, The great God, The true God,* and *if it shall please God*. This Expression in the Vulgar, is the Voice of Nature; and is it not also the Confession of Christians? And they who make *Jove* the Supreme Deity, mistake indeed in the Name, but agree in the Thing, in the Notion of One Almighty.

XIX. I find the Poets likewise singing of one Sovereign Deity, *Father of Gods and Men*, and who fashion'd our Souls according to his own Will and Pleasure. What says *Virgil* of *Mantua*? Does not he yet speak more plain, and nearest to Truth? *In the Beginning* (says he) *a Spirit quicken'd Heaven and Earth*, and all the Parts of the Universe, and *a Mind infus'd actuated the whole Mass, the Author of Men and Beasts*, and every Animal. The same Poet in another place calls this Mind and Spirit, God; his Words are these,

- - - - *Deum namque ire per omnes
Terrasq; tractusq; maris, cœlumq; profundum;
Unde homines & pecudes, unde imber & ignes.*

*Earth, Heaven, Sea, all Natures vast Abyss
Does God pervade and fill.
Hence Man, and Beast; Storm, and red Lightning hence.*

And what other God do we Christians preach up, than Mind, and Reason, and Spirit? Let us run over the Doctrine of the Philosophers, if you please, and you'll find them, though differently expressing themselves in Words, yet as to the Thing, all conspiring in one and the same Opinion. I omit the ancient Wise Men of *Greece*, deservedly so call'd upon the account of their Sayings. Let *Thales* the *Milesian*, the Principal of them, serve for the rest, who was the first that discours'd accurately concerning Heavenly Matters. This same *Milesian Thales* affirm'd Water to be the Principle of Things; but withal, that God was that Mind which form'd every Being out of this Fluid into a World. But let me tell you, this Account of Water and the Spirit was a Notion far above the reach of any Mortal, had not God, who was that Spirit, reveal'd it to *Moses*. Thus you see how the principal Philosopher exactly concurs with us Christians. After him *Anaximenes*, and then *Diogenes* sirnamed *Apolloniates*, make God of a Nature Aerial, Infinite, and Immense. These then agree with us in the Doctrine of One God. The God of *Anaxagoras*, is an Infinite Mind that disposes and puts every Thing in Motion; and the God of *Pythagoras*, is a Mind that permeates and takes care of the Universe, and is the Original of all Life. *Xenophanes* is well known to have defin'd God to be an Animated Infinite. *Antisthenes* holds a Plurality of Gods over several Nations, but one only Deity Supreme by Nature. *Speusippus* is for that natural, animal Virtue, by which All things are conducted, to be God. Does not *Democritus*, although

Gen. i. 2.

although the first Inventor of the Atomick Philosophy, does not he often speak of that Nature, which is the Author of those Images and Intentional Species, by which we understand, and feel, and see, &c. as God? *Strato* likewise sets up Nature for God. Even your famous *Epicurus*, who makes either Unactive, or No Gods, Deifies Nature. *Aristotle* is at Variance with himself: however he assigns one Sovereign Power; for one while he calls Mind God, another while the World he will have God, and then again he makes God above the World. *Heraclides* of *Pontus* reels about in the same manner; however, he ascribes a Divine Mind to God, or else to the World, or else makes a pure Divine Mind it self to be God. *Theophrastus*, and *Zeno*, and *Chrysippus*, and *Cleanthes*, though all at Difference, yet at the long run they all meet in the Notion of one Providence that superintends the World. For *Cleanthes* sometimes makes God to be a Mind, sometimes a Soul, sometimes *Æther*, and sometimes Reason. His Master *Zeno* makes the Natural and Divine Law in Things to be God; and sometimes *Æther*, and sometimes Reason, to be the Fountain of all. This same Philosopher, methinks, by calling *Juno* Air, *Jupiter* Heaven, *Neptune* Sea, *Vulcan* Fire, and all the rest likewise of the Popular Gods, Elements, by such Names severely confutes, and lashes the Publick Vanity of worshipping such Deities. *Chrysippus* says much the same Things; for he believes God sometimes to be a Divine Energy, a Rational Nature; and then again the World, and then fatal Necessity; and copies after *Zeno* by interpreting the Fables, of the Gods in the Verses of *Hesiod*, *Homer* and *Orpheus*, into Natural Principles. And *Diogenes* the *Babylonian*, expounds the lying-in of *Jove*, and the Birth of *Minerva*, and such like Fictions, not of the Gods, but of Nature. *Xenophon*, the Disciple of *Socrates*, affirms the Form of the true God to be Invisible, and therefore not to be search'd after. *Aristo* of *Chios* affirms him likewise to be Incomprehensible; and both the one and the other understood the Divine Majesty best, by despairing to understand it. But of all the Philosophers, *Plato* deliver'd himself the plainest and truest of God, of Things, and Names; and his Discourses had been purely Divine, had he not sometimes comply'd too far with the Vanities of the Age, and allay'd them with the Errors in Fashion. This same *Plato* therefore in his *Timæus* says, That by the Name *God*, we are to understand the Parent of the World, the Architect of the Soul, and the Maker of Heaven and Earth; whom it is hard to understand by reason of his incredible Immensity of Power, which is too much for Human Intellect; and when we do come to the Knowledge of Him, 'tis impossible to make our Notions intelligible to All. And we Christians almost say the same Things; for we are come to the Knowledge of this true God, and we also call him the Parent of all Things; nor do we preach these Divine Mysteries in Publick, but when the Publick calls us in Question about our Religion.

XX. I have now run over the Opinions almost of all the Philosophers, those of Note especially, whereby 'tis evident, they all declare

clare for one God, though under different Denominations; insomuch, that ev'ry one must conclude, either that the Christians now are Philophers, or that the Philosophers of old were Christians.

Lord Bacon.] I had rather believe all the Fables in the *Legend*, and the *Talmud*, and the *Alcoran*, than that this Universal Frame is without a Mind. And therefore God never wrought a Miracle to convince *Atheism*, because his ordinary Works convince it. It is true, that a little Philosophy inclineth Man's Mind to *Atheism*, but Depth in Philosophy bringeth Mens Minds about to *Religion*. For while the Mind of Man looketh upon Second Causes scattered, it may sometimes rest in them, and go no further: But when it beholdeth the Chain of them Confederate and Linked together, it must needs fly to *Providence* and *Deity*.

Essay on Atheism.

Mr. Boyle.] The Power and Wisdom of God display themselves by what he does, in reference both to his Corporeal and his Incorporeal Creatures. Among the manifold Effects of the *Divine Power*, my intended Brevity will allow me to mention only Two or Three, which tho' to discerning Eyes they be very manifest, are not wont to be very attentively reflected on. The *Immense Quantity* of Corporeal Substance, that the Divine Power provided for the framing of the Universe; and the *Great Force* of the Local Motion that was imparted to it, and is *regulated* in it. And first, the *Vastness* of that huge Mass of Matter that this Corporeal World consists of, cannot but appear stupendious to those that skilfully contemplate it. That part of the Universe which has been already discovered by Human Eyes, assisted with Dioptrical Glasses, is almost uncon-

Of the Veneration Man's Intellect owes to God, p. 10.

unconceivably vast, as will be easily granted, if we assent to what the best Astronomers, as well Modern as Ancient, scruple not to deliver. [See Pag. 56, 57. above,] And it plainly appears by the Parallaxes, and other Proofs, that this Globe of Earth and Water that we Inhabit, and often call the *World*, though it be divided into so many great Empires and Kingdoms, is so far from being for its Bulk a considerable part of the Universe; that, without much Hyperbole, we may say, that 'tis in Comparison thereof, but a *Physical Point*. Nay, those far greater Globes of the Sun, and other fixed Stars, and all the solid Masses of the World to boot, if they were reduc'd into One, would perhaps bear a less Proportion to the Fluid [empty] part of the Universe, than a *Nut* to the *Ocean*. Which brings into my Mind the Sentence of an excellent modern Astronomer, That the Stars of the Sky, if they were crouded into One Body, and placed where the Earth is, would, if that Globe were placed at a fit Distance, appear no bigger than a Star of the first Magnitude now does. And after all this, I must remind you, that I have been hitherto speaking but of that part of the Corporeal Universe that has been already seen by us. And therefore I must add, that as vast as this is, yet all that the Eye, even when powerfully promoted by prospective Tubes, hath discovered to us, is far from representing the World of so great an Extent, as I doubt not but more perfect Telescopes hereafter will do. ----From the vast *Extent* of the Universe, I now proceed to consider, the stupendious *Quantity of Local Motion*, that the Divine Power has given the Parts of it, and continually maintains in it.

[See

p. 11.

p. 12.

p 13.

p. 14.

of Religion. 233

[See Pag. 53, 54. *prius*,] ---These Things are men- p. 21.
tion'd, that we may have the more enlarg'd Conceptions of the *Power*, as well as *Wisdom* of the
Great Creator, who has put so wonderful a
Quantity of Motion into the Universal Matter,
and maintains it therein; and is able not only to
set Bounds to the raging Sea, and effectually
say to it, *Hitherto shalt thou come, and no far-* Job
ther; and here shall thy proud Waves be stayed: xxxviii.
But (what is far more) so [by the Power of 11.
Gravity] to curb and moderate those stupen-
diously rapid Motions of the Mundane Globes,
and *intercurrent Fluids*, [rather, *in the inter-
pos'd Vacuities*] that neither the Unweil-
dinefs of their Bulk, nor Celerity of their
Motions have made them exorbitate, or fly out,
and this for many Ages; during which, no
Watch for a few Hours has gone so regularly.
---The Contrivance of every Animal, and espe- p. 24.
cially of a Human Body, is so curious and exquisite, that 'tis almost impossible for any
body that has not seen a Dissection well made,
and Anatomically consider'd, to imagine or
conceive how such excellent Workmanship is
display'd in that *admirable Engine.* ---I shall
here tell you in a word, (and 'tis no Hyperbole,)
that as St. *Paul* said on another Occasion, *That* 1 Cor. i.
the Foolish Things of God are wiser than Men; 25.
and the weak Things of God stronger than Men:
So we may say, that the meanest Living Creatures of God's making, are far more wisely
contriv'd than the most excellent pieces of
Workmanship that Human Heads and Hands can p. 25.
boast of. And no *Watch* nor *Clock* in the World
is any way comparable for Exquisiteness of Mechanism to the Body of even an *Ass* or a *Frog*.
---We need not fly to Imaginary ultramundane p. 43.
Spaces,

Spaces to be convinc'd that the Effects of the *Power* and *Wisdom* of God are worthy of their Causes, and not near adequately understood by us; if, with sufficient Attention, we consider that *innumerable Multitude*, and *unspeakable Variety* of Bodies that make up this *vast Universe*. For there being among these a stupendious Number that may justly be look'd upon as so many distinct Engines, and many of them very complicated ones too, as containing sundry subordinate ones; to know that all these, as well as the rest of the Mundane Matter, are every Moment sustain'd, guided, and govern'd according to their respective Natures, and with an exact Regard to the Catholick Laws of the Universe; to know, I say, that there is a *Being* that doth this every-where, and every Moment, and that manages all Things without either Aberration or Intermission; is a Thing that, if we attentively reflect on, ought to produce in us, for that Supreme Being that can do this, the *highest Wonder*, and the *lowliest Adoration*. [*See the rest of that excellent Discourse.*]

Dioptricks. p. 195.

Mr. *Molyneux.*] I should think it an Attempt worth the Thought of some profound Philosopher, to give an Account of those Admirable, Orderly, and Beautiful Appearances of Nature, whereof we can most plainly apprehend the *Designs*, and *final Causes*, but can hardly proceed to any further Knowledge of them. ---This surely might be able to convince the most obstinate Opposers of *Divinity*. For certainly, if we can rely upon *any Deduction*, or *Consequence* drawn out by the *Mind* of *Man*, we may assuredly rest satisfied in this, That so many *Phænomena*, stupendous and surprizing, for their
design'd

of RELIGION.

design'd Contrivance, could not proceed but from an *Omnipotent* and *Designing* Being.

And from hence may we justly fall into the deepest Admiration, that *one* and the *same Law* of Motion should be observed in Bodies so vastly distant from each other, and which seem to have no Dependance or Correspondence with each other. This does most evidently demonstrate, that they were all at first put into Motion by *one* and the *same unerring Hand*, even the *infinite Power* and *Wisdom* of God, who has *fix'd* this *Order* among them *all*, and has *established* a *Law*, which they cannot *transgress*. *Chance* or *dull Matter* could never produce such an *Harmonious Regularity* in the Motion of Bodies so vastly distant: This plainly shews a *Design* and *Intention* in the *first Mover*. And, with Submission to the Reverend and Learned *Divines*, I am apt to think that one Argument drawn from the *Order, Beauty* and *Design* of Things, is more forcible against *Atheism*, than Multitudes of Notional Proofs drawn from *Ideas*, Apparitions of *Spectres*, *Witches*, &c. (not that these should lose their due Strength) For besides the *Heavens*, even the *little Globe* we inhabit affords us infinite Variety in this Kind: And for my own part, I must confess, I can read more Divinity in Mr. *Charlton*'s admirable *Musæum*, on a Box of beautiful *Shells*, of delicately Painted *Plants*, curiously adorned *Insects, Serpents, Birds*, or *Minerals*; than in large Volumes of Notional Writers. For *Animals, Plants*, and *Minerals*, do yield us abundant Instances, which visibly shew a *Design* or *End proposed*; which, as it cannot possibly consist with *Chance*, so neither can

p. 273, 274.

can it be apprehended to have been so *ab æterno*: For 'tis absolutely unconceivable, that a Thing *designed* for some *End* or *Purpose*, should not be so *designed* in *Time*, by some *designing Being*.

Optic. Ed. Lat. p. 314, 315.

Sir *Isaac Newton*.] It is the Principal Thing that Natural Philosophy ought to do, and the End of that Science, that by a Chain of Reasoning, we proceed from Effects to their Causes, until we arrive at the very *First Cause* it self. That we do not only explain the Mechanism of the World; but that besides this, and as the Fruit of our Enquiries about it, we answer these following *Queries*, with others of a like Nature: What there is in the Celestial Spaces void of Matter? And whence it is that the Sun and Planets gravitate mutually towards one another, while the Spaces between are void of Matter? How it comes to pass that Nature acts nothing in vain? And whence proceeds the admirable Beauty of the Universe? To what End the Comets were made? And whence it is that they move in Orbits so very Eccentrical, from and to all Parts of the Heavens? whereas the Course of the Planets hath the same Direction, towards the same Parts, in Orbits Concentrical? And what hinders the Sun and fixed Stars from rushing mutually upon one another? How it comes about that the Bodies of Animals are fram'd with such exquisite Art and Wisdom? And for what Purposes their different Parts were fitted? Whether it were possible that the Eye could be framed without the Knowledge of Opticks? Or the Ear without the Knowledge of Sounds? Whence it is that the Motions of the Body obey the command of the Will? And whence

whence is what we call *Instinct* in Animals? Whether the Sensory of Animals be not the Place where the Substance which has Sensation is present, and into which the sensible Species of Objects are carried by the Nerves and the Brain, that they may there be perceived where they are actually present by that Substance there present? And whether from a right Solution of these Queries, it does not appear that there is a Being, Incorporeal, Living, Intelligent, Omnipresent, who in infinite Space, as it were in his Sensory, sees accurately and intimately, and discerns throughly the Things themselves; and by being present to them comprehends them all within himself: Of which Things, that which in us Perceives and Thinks, Perceives and Beholds in its little Sensory, only the Images, brought to it by the Organs of Sense?

This most excellently contrived System of the Sun, and Planets, and Comets, could not have its Origin from any other than from the wise Conduct and Dominion of an Intelligent and Powerful Being. And in case the Fixed Stars be the Centers of the like Systems, they that are framed by the like wise Conduct, must all be subject to the Dominion of *One Being*; especially while it appears that the Light of the Fixed Stars is of the same Nature with the Light of the Sun; and that all these Systems do mutually impart their Light to one another. *Philos. Natural. Princip. Math. 2 Edit. Scholium Generale, p. 482, 483.*

This Being governs all Things; not as a Mundane Soul, but as the Lord of all Creatures; who on account of his Dominion over them, is usually stiled the *Lord God*, παντοκράτωρ or *Supreme Governor of the Universe*. For the Word *God* is relative, and hath Relation to subordinate Beings:

And

And the Word *Deity* imports the Exercise of that Dominion, not over his own Body, (as is the Opinion of those that make him the Soul of the World) but over those subordinate Beings. The *Supreme* God is an Eternal, Infinite, and Absolutely Perfect Being: But a Being that is never so Perfect, is not a *Lord God* without Dominion. For we say, *My God, Your God, the God of* Israel: But we don't say, *My Eternal, Your Eternal, The Eternal of* Israel: We don't say, *My Infinite, Your Infinite, The Infinite of* Israel: We don't say, *My Perfect, Your Perfect, The Perfect of* Israel: These Denominations of him having no Relation to subordinate Beings. The Word *God* most frequently signifies *Lord*; but so that every Lord is not a God. The Exercise of Dominion in a Spiritual Being, constitutes a God: If that Dominion be Real, that Being is a Real God; if it be Supreme, the Supreme God; if it be fictitious, a false God. And the Consequence of the Exercise of real Dominion, by the true God, is this, That He is a Living, Intelligent, and Powerful Being; as it is the Consequence of the rest of his Perfections, that he is the Highest or most perfect Being. He is *Eternal*, and *Infinite*, and *Omnipotent*, and *Omniscient:* That is, he endures from Everlasting to Everlasting, and is present from Infinity to Infinity; He governs all Things, and knows all Things that are done, or can be known. He is not Eternity, or Infinity, but an Eternal, and an Infinite Being. He is not Duration, or Space, but he is a Being that Endures, and is Present: He endures always, and is present every where; and by existing always and every where, he constitutes Duration and Space, Eternity and Infinity. Since every Particle of Space exists *always*,

ways, and every indivisible Moment of Duration exists *every where*, 'tis evident the Framer and Lord of all Things cannot exist *never* or *no where*. He is Omnipresent, not only by his *Power*, but also by his *Substance:* For Power cannot subsist without Substance. All Things are † contained and move in him, but without his Suffering thereby. God suffers nothing by the Motions of Bodies, nor do they feel any Resistance by the Omnipresence of God. 'Tis well known, that the Su-

---

† This was the Opinion of the Ancients: Aratus] *Let us begin with Jove: Let us Men never leave off discoursing of him: For every Concourse of People, every Assembly of Mankind, the Seas also, and the Heavens are all full of Jove. We all enjoy the Blessings of Jove: For we are also his Offspring.* Phænom. at the Beginning. Paul] *That they should seek the Lord, if haply they might feel after him, and find him; though he be not far from every one of us. For in him we live, and move, and have our Being; as certain also of your own Poets have said; For we are also his Off-spring,* Acts xvi. 27, 28. Moses.] *Know therefore this Day, and consider it in thine Heart, That the Lord he is God, in Heaven above, and in the Earth beneath; there is none else,* Deut. iv. 39. *Behold the Heaven, and the Heaven of Heavens is the Lord's thy God; the Earth also, with all that therein is,* X. 14. David.] *Whither shall I go from thy Spirit? Or whither shall I flee from thy Presence? If I ascend up into Heaven, thou art there. If I make my Bed in Hell, behold thou art there.* Psal. cxxxix. 7, 8. Solomon.] *Will God indeed dwell on the Earth? Behold the Heaven, and Heaven of Heavens cannot contain thee; how much less this House that I have builded?* 1 King viii. 27. Job.] *Is not God in the Height of Heaven? And behold the Height of the Stars how high they are!* xxii. 12. Jeremiah the Prophet] *Am I a God at Hand, saith the Lord, and not a God afar off? Can any hide himself in secret Places, that I shall not see him, saith the Lord? Do not I fill Heaven and Earth, saith the Lord?* xxiii. 23, 24.

preme God exists of Necessity; and by the same Necessity does he exist *always* and *every where*. Whence it is that he is entirely like himself, all Eye, all Ear, all Brain, all Arm, all Sensation, all Intelligence; all Action; but this in a way not at all like Men; in a way not at all like Bodies, in a way utterly unknown to us. As a Blind Man has no Idea of Colours, so have not we any Idea of the *Modus*, whereby God, most wise, perceives and understands all Things. He is entirely void of all Body and Bodily Figure; and therefore cannot be either seen, or heard, or felt; nor ought he to be worshipp'd under any Bodily Shape. We have the Idea's of his Attributes, but do not at all know what the Substance of any Thing is. We see only the Figures and Colours of Bodies, we hear only their Sounds, we feel only their outward Surfaces, we smell only their Scents, and we taste only their Savours; but we don't know their inmost Substances by any Sensation, or internal Reflection; and much less have we any Idea of the Substance of God. We Know him only by his Properties, and Attributes, and the most wise and excellent Structures of his Creatures, and by final Causes; while we Adore and Worship him on Account of his Dominion. For a God, without Dominion, Providence, and Final Causes, is nothing else but Fate and Nature. And thus much concerning God; To discourse of whom, from the Appearances of Nature, does certainly belong to *Experimental Philosophy*.

PART

PART IX.

A Recapitulation *of the Whole:* With *a serious* Address *to All,* especially *to the* Scepticks *and* Unbelievers *of our* Age.

AND now, Reader, whofoever thou art, efpecially if thou beeft a Sceptick, or Unbeliever, either as to Natural or Revealed Religion, I beg of thee ferioufly to look back upon what has been hitherto Difcourfed on the Behalf of them both; even from the certain Principles of Aftronomy, or the true Syftem of the World; and from thofe numerous Teftimonies of Sacred and Prophane Antiquity, which fhew us the natural Confequences of fuch wonderful Phænomena. I fay, look back *ferioufly* upon this View of the Univerfe before us, and its Confequences. For if ever there be Occafion for *Serioufnefs*, it is here, where our *All* is at Stake; where our future, our final Weal, or Woe, Happinefs or Mifery,

R are

are the Things under Examination. For accordingly, as we shall determine our selves in this grand Enquiry, concerning the Being and Providence of God, the Immortality of our Souls, and the Truth of Divine Revelation, as to the lasting Rewards and Punishments of another World; so shall we be oblig'd to behave our selves in our Conduct; upon which our Eternal State is to be awarded us at the great Day. For we cannot but be sensible that no Mistake of our own can alter the Nature of Things; and that they are not the most zealous Wishes, and Inclinations; the most pungent Jests and Banter; the most Prophane and Impious Blasphemy against God, his Attributes, or Providence, that can in the least alter the System of the Universe, or banish the Supreme Creator and Governor, with his Providence and Laws, out of it. Let us consider, then, that all the other Hypotheses relating to the Constitution of the World, invented by either *Democritus*, *Epicurus*, *Aristotle*, *Ptolemy*, *Tycho*, *Cartes*, Mr. *Hobbs*, or *Spinoza*, do now plainly appear, from certain Evidence, to be not only false, but absurd; contrary both to common Sense, and to the known Laws and Observations of sound Philosophy; and that he who will now be an *Atheist*, must be an absolute *Ignoramus* in Natural Knowledge; must neither understand the Principles either of Physicks or Astronomy. Let us consider farther, that as to *Deism*, or the Denial of the Scriptures, and of Divine Revelation, it is really Ill Mens *last Refuge*, and taken up of late, not by honest Enquirers, impartially searching after Truth, and discovering upon Evidence, that
all

all Revealed Religion is false; but that it is chiefly fallen into of late, by some Irreligious Persons, in the Distress of their Affairs, and upon that surprizing and overbearing Light, which Sir *Isaac Newton*'s wonderful Discoveries have afforded; whereby they have perceiv'd that Natural Religion, with its Foundations, were now become too certain to bear any farther Opposition. That this is true, I appeal to a certain Club of Persons, not over-religiously dispos'd, who being soberly asked, after Dr. *Bentley*'s remarkable Sermons at Mr. *Boyles*'s Lectures, built upon Sir *Isaac Newton*'s Discoveries, and level'd against the prevailing Atheism of the Age, *What they had to say in their own Vindication against the Evidence produc'd by Dr.* Bentley? The Answer was, *That truly they did not well know what to say against it, upon the Head of Atheism: But what,* say they, *is this, to the Fable of Jesus Christ?* And in Confirmation of this Account, it may, I believe, be justly observ'd, that the present gross *Deism*, or the Opposition that has of late so evidently and barefacedly appear'd against Divine Revelation, and the Holy Scriptures, has taken its Date in some Measure from that Time. And as to the main Observation which I am now upon, I mean that this modern Infidelity is not properly owing to any new Discovery of the want of real Evidence for Reveal'd Religion, or of the Falsity of any of the known Foundations of it; but to the like Necessity of Affairs, and the Impossibility of supporting the former, and worser Notions, I think is plain from these Two farther Considerations: First, That the most truly Learned, the deepest Enquirers, and most Sagacious Examiners into Reveal'd Religion, have in this Age, as well as in all the foregoing, declar'd themselves in Favour of it, both by their Conduct, and by their Writings; while the Generality of the *Deists* are known to be so Overly and Superficial in their Learning, about such Matters, as renders them indeed sometimes the Scare-crows of the Ignorant, but generally the Contempt of the really Judicious and Learned Christians. Secondly, That they have, for some Time, almost discarded the principal way of Examination into the main Evidence for the *Jewish* and Christian Revelations, I mean Ancient Facts and Testimonies; which they would never have done, had they not been well assur'd that such Sort of Arguments would not be for their Service. However, I shall wave this, as somewhat Foreign to my present Undertaking, and proceed to that *Recapitulation* of what I have alledg'd in this Treatise, on the Behalf of the Being, and the Attributes, and the Providence of God, of the Immortality of Human Souls, and of the several important Points of Natural and Reveal'd Religion already treated of, which I proposed to make in this Place.

We have then here, Good Reader, seen a wonderful, a surprizing, an amazing System, or rather an innumerable Number of such Systems of Worlds; *i. e.* of Suns, of Planets, Primary and Secondary, and of Comets, with their several Atmospheres, all placed at immense Distances from one another; in various Positions, Velocities, and Periods; in divers Circumstances and Magnitudes, ordain'd for several great Uses, and admitting different particular Laws; but so as every where to be subject to

one

of Religion.

one Universal *Power of Gravity*, or mutual Tendency of all the Parts to one another, and that of a certain Quantity, and in certain Proportions. A Power this, amazing to think of! yet Undeniable, Regular, Universal as to Time, Place, and Bodies; and still Exact and Geometrical; yet at the same Time entirely and absolutely Immechanical, or beyond all Material Solutions, and Pretence of Material Solutions whatsoever; and indeed the proper Effect of a Supreme Being. We have seen, that accordingly this Immense World, or Innumerable Number of Immense Worlds, are for certain all *God's Worlds*; or Created, Governed, and Provided for by One God, by One Supreme, Omnipotent and Omniscient Being; ever Present to all its Parts, and ever exerting his Infinite Power, Wisdom, and Goodness every where therein. We have seen that the noblest Principles of Natural Religion are Fully and Demonstratively deriv'd from the Phænomena of these Systems; and that not a few of the most Concerning, and otherwise most Exceptionable Parts of Divine Revelation, are also strongly confirm'd thereby. Nor is there now the least room for either of those Ancient Refuges of Atheism and Irreligion; I mean the wild Hypotheses of the *Eternity of the World*; and of its Temporary Derivation from the *Accidental Concourse of Atoms*. All this we have now *seen* with our Eyes, and, as it were, *felt* with our Hands, in the foregoing Treatise. We have also there observ'd and prov'd, that this wonderful System of Things is not any bare Hypothesis, or meerly *probable Account of* the Heavenly Bodies, and their Motions, but

but the *certain* Theory of them, attefted to be fuch by unqueftionable Evidence, from Aftronomical Obfervations, and from fure Geometrical Reafonings thereupon. So that the Conclufions regularly drawn from fuch Premifes, ought themfelves to be look'd on as *Certain*. We have there alfo diftinctly followed the Steps of Nature, and drawn out her grand Secrets into plain Tables, for the Ufe of every body; even of thofe who are not Mathematicians good enough to Calculate themfelves: And have farther given the Manner and Reafon of each Operation, for every ones entire Satisfaction. Nor need my Readers take even the *Lemmata* themfelves for granted, if they underftand but fome Elements of Geometry. For they are demonftrated every one in my *Mathematical Philofophy*, and that generally after fo plain a Manner, that I dare fay very ordinary Mathematicians will be able to underftand thofe Demonftrations. By thefe Calculations it is that we arrive at the compleateft and moft exact Knowledge of this *Noble*, this *Amazing*, this *Divine Syftem*. Nor can I imagine that the Inquifitive Reader, when he has well confider'd the Particulars, will think that any of thofe Epithets, *Noble*, *Amazing*, and *Divine*, are by me wrong apply'd in this Matter. For here we difcover that all the Heavenly Bodies *Revolve* in thofe moft agreeable Geometrical Curves, the *Ellipfes*; the Planets in thofe that are very little Eccentrical; and the Comets in thofe that are prodigioufly fo; even in fuch as are almoft Parabolical. By which known and Regular Orbits, we readily reduce their Motions to Calculation, and eafily difco-

of RELIGION.   247

discover the Law of Gravity belonging to them. Here we contemplate the *Periods* of all the Planets Primary and Secondary, and of the Comets, about their Central Bodies, from the smallest Period of the Innermost Circumjovial of 42 Hours and a half; to the largest of the outmost Comet of 575 Years. Here we take a View of the vast middle *Distances* of all the Planets, Primary, and Secondary, and of the Comets, from their Central Bodies, deriv'd from the best Observations, from the least of the innermost Circumjovials of 130.000, to the greatest of the utmost Comet of 5.600,000.000 Miles; the least of which distances does prodigiously surpass the Power of Human Imagination, which can no way enlarge it self to any such measures. Here also we may discover the surprizing *Magnitudes* of the several Bodies belonging to our System, both in Diameter, Superficies, and Solidity, from the smaller Quantities in our Moon, of 2170 Miles Diameter; 14,000.000 square Miles of Surface; and 5.000,000.000 cubical Miles, of Solidity; to the vastly greater Quantities in the Sun of 763.000 Miles Diameter; 1,813.200,000.000 square Miles of Surface; and 230.000,000.000,000.000 cubical Miles of Solidity. Numbers that are still more immensely beyond all Human Imagination; and, such as, if *Epicurus* or *Lucretius* were alive, who could raise their stupid Conceptions no farther than the largeness of a Cart-wheel for the Sun it self, would have quite affrighted them out of their foolish Philosophy. We here learn the *Annual Velocities* of the several Planets, Primary and Secondary; and of the Comets, from the slowest Motion in the Moon, of 2200 Miles, to the swiftest in *Mercury*, of 100.000 Miles, in the space of one Hour. As also we here learn the *Diurnal-Velocities* at the Equator, the slowest of the Moon of 10 Miles, the swiftest of *Jupiter* of 25.000 Miles in the same space of one hour: And that even we upon the Surface of this Earth, when we cross the Line, with all our Buildings, move along 1030 Miles in an Hour; and that by Consequence, in this Latitude, I my self, while I seem to be at rest, writing this Treatise in my Study, do yet, together with my Study, and my Books, revolve at the rate of above 600 Miles in the same time. Nay, we here get a step farther, and, without going down into the Central Regions of any one of the Celestial Bodies, do certainly pronounce, not only concerning the Mathematical *Quantity* of Bulk or *Magnitude*, but in many Cases concerning the real Physical *Quantity of Matter* contain'd in those Bodies, as compar'd, I mean, with one another; which is as far as Philosophy can possibly carry us. Whereby we find that the Moon, which is the least of all those Bodies whose Quantities of Matter we know, is not quite the 9,000.000th part so great in this respect as the Sun; and that the Quantity of Matter of all the Planets and Comets taken together, does not in probability amount to the 500th part of that in the Sun alone; the *Sun*, I say, that most amazing and most prodigious Creature of God that is in this System, and perhaps in all the visible Systems about us! of which Mr. *Milton* bravely sings;

Paradise Lost. Lib. V.

Thou Sun, of this great World, both Eye and Soul, Acknowledge God thy Greater : Sound his Praise In thy Eternal Course!

More-

of RELIGION. 249

Moreover, we here certainly discover by Consequence, the very inward Texture of the same Celestial Bodies, whose Quantities of Matter were above determin'd; and, without digging into the Bowels of any of the Planets, do, with equal Certainty, pronounce what comparative degree of *Density* they have: Whereby we learn that the Sun is vastly too dense for a Flame, as *Cartes* determined; and that the Moon is the Densest, and *Saturn* the Rarest of all these Bodies; and that the former is above eight times as dense as the latter. Which Conclusions shew the admirable Nature, and profound Reach of Sir *Isaac Newton*'s Philosophy, which, with equal ease and certainty, penetrates to such deep Truths, as no other Hypotheses do so much as in the least hope or pretend to attain to, even by Conjecture. Nor do we stop here; but, placing our selves on the several *Surfaces* of the fore-mentioned Planets, we, by certain Reasoning, determine the comparative *Weight* of any given Body on those several Surfaces; and observe, with great Satisfaction, that the same Strength that can here lift One Hundred Pound, would not be able on the Sun's Surface to lift Four Pounds; and that what on the Moon's Surface, where this Weight is the least, would weigh one Pound; would, if transferr'd to the Sun's Surface, where it is the greatest, weigh above 71 of the same Pounds. We do also here Contemplate those *Diurnal Motions* with respect to the other Heavenly Bodies, which we are forced to gather by Geometrical Reasoning, with respect to our own Earth; and which the several Inhabitants of those Bodies, (if such they are, and if they be provided with

such

such Means of seeing our Earth, as our Telescopes have of late afforded us for seeing them) may also Contemplate with their Eyes. Whereby we find that they do revolve in Periods sufficiently unequal; from the slowest, that of the Moon, in 27 Days, to the swiftest, that of *Jupiter*, in a little less than 10 Hours: which Sight seems to me to be a kind of sensible Confirmation of the like Diurnal Motion of our Earth.

We are here also taught to estimate the different *Degrees of Heat and Light*, which our Fellow Creatures derive from the great Fountain of both, the Sun, when nearer, and when farther off than we are: Which difference is so vastly great, even at the mean Distances of the several Bodies from the Sun, that *Mercury*, the nearest, has in a Mean no less than 120000 times the Quantity of Light and Heat which the utmost Comet has; and that the Heat at the Sun's Surface is no less than 45000 times as great as any part of this Earth receives from it, at any time; and that withal the outmost of our known Comets abides at one time an Heat more than 400,000.000 of times, as great as it does another. An *amazing difference* this! and such as is no where else to be parallel'd, that we know of, in the whole System of Nature. We here also see, how soon our Earth, and any of the Planets would *fall to the Sun*, or to their central Bodies, if their Projectile Velocities should cease; from the innermost of the Circumjovials, which would fall to *Jupiter* in 7 Hours; to the outmost Comet which would not fall from its middle distance to the Sun under 66 Years. We have also taken a

view

of RELIGION. 251

view of our neighbouring Body the Moon, and found it in almost all respects such a Planet, or Place of Habitation, as our own Planet the Earth is. And we should be greatly wanting to our Selves, and to the *Decorum* of Things, as well as highly injurious to our great Creator, if we should so much as scruple the Supposition of so noble an Habitation's being Inhabited; of so noble a Colony's being Peopled. Those who can attentively view the Wonderful, and Beautiful and Admirably contriv'd Structure of this our adjoining Planet with its Sea and Land, Mountains and Valleys, Day and Night, Summer and Winter; together with its Clouds and Atmosphere, and Moon; all in correspondence to our Earth, which is every where full of intelligent Beings its Inhabitants; and yet shall peevishly deny that it either now is, or ever was, or is to be in like manner inhabited by such Intelligent Beings; and are resolv'd it shall have no other Use than to enlighten our Earth, and be peep'd at through our Telescopes; seem to me too Unphilosophical to be Argued with; and only worthy to be left to their own narrow Genius, which can neither think a brave uncommon Thought, nor admit any thing but what their Education or System have already forc'd upon them. Nor indeed, do we need to debate here with such Men; because it can hardly be imagined they will ever have Skill or Curiosity enough to peruse, what either has already been written, or may hereafter be written upon such great and noble Subjects. Besides this single Attendant of ours, we have taken a Prospect of Four such Attendants on *Jupiter*, and Five on *Saturn*, with the remarkable *Bells*, and

and Diurnal Revolution of the former; and the much more aſtoniſhing *Ring* of the latter: The One bearing a near reſemblance to our Clouds under the Torrid Zone, upon our Earth's daily Motion, and ſo hinting to us the Likeneſs there is between that and our own Planet: And the Other affording an Inſtance of a more ſurprizing Variety in the Works of God and Nature than is any where elſe to be ſo evidently diſcover'd in the World. Nor muſt I here paſs over in Silence that wonderful, that prodigious, that amazing Inſtance of *Swiftneſs of Motion*, which the Rays of Light afford us, and which is gathered from the Eclipſes of *Jupiter*'s Planets; no leſs I mean than that of 180.000. Miles in one Second of Time; whence it appears, that a Being, might viſit all the Men in the World, if he proceeded with the Swiftneſs of theſe Rays, and thoſe Men were placed in any tolerable Order along or near a great Circle of the Earth, in a very few Seconds of time. Nor do any of the other ſwifteſt Motions that we know of in the Univerſe deſerve to be called *ſwift* in compariſon of this before us; which is no other than a conſtant Inſtance of the Power of God in moving the Bodies which he has made, with what Velocity he pleaſes; and thereby of communicating his Influences, even in a Mechanical Way, to immenſe Diſtances, almoſt in a Moment. But then, beſides this Planetary World, which was in ſome meaſure known to the Ancient Aſtronomers, we have alſo taken a View of another World, or Species of Bodies, known indeed by Name, but hardly at all by Nature to the Ancients; I mean the Syſtem of Comets, which till lately were generally

of RELIGION.

rally look'd on as inconsiderable and fortuitous Meteors of our Air only; but do now appear to be more numerous, and not less considerable than the Planets themselves. These Comets pass so entirely through the Planetary Regions, and may so certainly approach to the Planets themselves both in their Descent and Ascent; that they may serve hitherto unknown Ends of Providence, both in their own Constitution, and by the Changes they may occasion in the Planets; and do therefore well deserve our most attentive Consideration. Nor certainly were the Planets, their Number, Magnitude, Motions, and Uses so well known in some Thousands of Years after their first Observation, as the Comets now are in a few Hundreds, or rather Decads of Years since they were to any good Purpose observed by Astronomers. And then, lastly, after all, we have taken a short imperfect View of the vastly numerous, the vastly great, and vastly distant Systems of the Fixed Stars, or to us new Systems of Worlds quite remote from this our Planetary and Cometary World: In comparison of all which Systems of Worlds, our own entire System, with its Sun, and all its Planets and Comets, must be but inconsiderable; probably not the $10{,}000^{th}$, perhaps not the $100{,}000^{th}$, or $1{,}000{,}000^{th}$ Part of the Whole: And whose Distance appears to be so great from us, that a Musket Bullet that should go 240 Feet in one Second of Time, if it had been shot up at the *Mosaick* Creation to the nearest fixed Star, and continued its Course evenly all the Way, would hardly have arrived there by this time, after the long Interval of 5700 Years. So immensly numerous, and immensly great and glorious is the entire System; and so inconsiderable

rable are we poor Worms, creeping upon this little little Earth, if compared thereto; even *as Nothing, yea less than Nothing, and Vanity!* And here we have plainly lost our selves in the amazing *Length* and *Breadth*, and *Heighth* of the Grand System, and of that Power, Wisdom, and Goodness, which shines forth in every Branch thereof; those chiefly excepted which depend on the Actions of Free-Creatures, and the manner of their Treatment by that Providence which discovers it self in the whole Universe; the Rules of whose Conduct are not yet laid fully open to our present Curiosity, but are rather reserved for the last and noblest Scene of our Duration hereafter. Nor is this to be so much wondred at, if we consider that the most beautiful, and orderly, and wisely contrived System of this visible World it self, which we have been just reviewing, tho' it has all along, since the beginning of the World, afforded many and noble Indications of that Divine Power, Wisdom, and Goodness to all Mankind, yet have the entire Secrets of that Contrivance, with that universal Power of Gravitation, by which the whole Machine has all along been upheld, lain in a manner hid through all past Ages, and is but just now discovered to us. Nay, the very best System of the Heavens, which the Astronomers long had, is reported to have appeared to One of them so aukward, absurd, and disagreeable, that he was not able to restrain his Tongue from a kind of Blasphemy against its Author; meerly because the grand Mysteries of the whole Machine, by the Knowledge of which all those seeming Disorders are now entirely vanished,

Isa. xl. 17.

Alphonsus.

of RELIGION.

nifhed, was not at that time difcovered to Mankind, but was referved for thefe later, and on that Account happier Ages of the World.

And now, Good Readers, having made this Review of the entire Univerfe, let us, in Agreement with the reft of my Defign, turn our Eyes from the Works to the Workman; from the Effects to the Caufe; from the Creatures to the Creator; from thefe Glorious, thefe Divine Works of Nature, to the more Glorious and more Divine Author of Nature, the great *God, Bleffed for ever*. For as it is excellently obferv'd in the Book of Wifdom, *Surely vain are all Men by Nature who are ignorant of God; and could not, out of the good Things that are feen, know him that is; neither by confidering the Works did they acknowledge the Workmafter.* — *For by the Greatnefs and Beauty of the Creatures proportionably the Maker of them is feen.* And if there be any Deductions of Human Reafon which are eafier and more obvious than the reft, this Way of Arguing, which we have already ufed, from the *Houfe* to the *Architect*; from the *Clock* to the *Clock-maker*; from the *Ship* to the *Shipbuilder*; and from a *noble, large, well-contriv'd*, and *well-proportion'd*, and *moft beautiful* Houfe, or Clock, or Ship, to the *excellent* Architect, the *skilful* Clockmaker, the *fagacious* Shipbuilder; this is fuch clear, natural, obvious, fure Reafoning, that we even at firft make ufe of it in Childhood, and find it as clear, natural, obvious, and fure in our elder Age; without occafion for a Tutor to inftruct us in it at firft, or for a Logician to improve us in it afterward. And fhall we reject that

xiii. 1.

v. 5.

that way of Reasoning, in the most eminent of all Instances, which we are not able to avoid making in the smallest? Shall the comparatively few, trifling, imperfect Contrivances of every small Machine here (which yet only applies the Powers of Nature to particular Purposes,) be universally, without Hesitation, allowed to prove a subtle, a shrewd, and a wise Contriver thereof? And yet, shall the numberless, the important, and the most compleat Contrivances, which surround us every where in this Universe, from the immensly great Body of a Sun, to the as prodigiously small Bodies of some *Animalcula*, be ascrib'd to *Fate*, to *Chance*, to any Thing imaginable, besides the great *Creator* and *Contriver* of all Things himself; to whom yet from the earliest to the latest Records of Mankind, as we have seen, the Wisest and Best have ever freely and unanimously ascribed them? But why do I speak of the *Wisest* and *Best* only in this Case? As if the rest of Mankind have generally had other Notions. No, the whole Race of Mankind, abating a very few, little better either for Ignorance, or Vice, or both, than Monsters, have still from one Generation to another drawn the very same Conclusions and Truths in their Minds; tho' they have not all made equal Application of those Conclusions and Truths to their Practice. In witness whereof, I might alledge the Concessions of not a few such bad Men; but shall chuse only to instance in the late famous Earl of *Rochester*, who long wanted, not so much Abilities to discover, as Goodness to make use of such Arguments. This Person therefore, as I have been informed, having been one Night deeply engag'd in Atheistical and Blasphemous Discourse among maong his Companions, as he too frequently was, after a while happen'd to have occasion to step abroad: where the Sky, being very clear, presented him with a glorious Prospect of no small Part of that beautiful World which we have been more distinctly describing. Upon the View of which he was overheard to say, *What a Dog am I, thus to blaspheme Him that made me and all this beautiful World!* I do not at present recollect from whom I had this Account; tho' the Thing it self was too remarkable to be forgotten. But whether any Mistake might be made in the Circumstances of this Story or not, 'tis unquestionable that such must frequently be the natural Reflections of a considering Mind, in these Circumstances; whose Impressions nothing can entirely supersede. As for my self, I must freely own, that as I had from my Childhood ever learned from the Works of God to acknowledge and worship Him that made them; and as I improv'd in Anatomy, in Astronomy, in Natural Philosophy, I saw that this first Impression or Voice of Nature was still more and more confirm'd and establifhed by farther Enquiries; so that when, in my younger Days, I had with great Difficulty and Pains, attained to the Knowledge of the true System of the World, and of Sir *Isaac Newton*'s wonderful Discoveries thereto relating, I was not only fully *convinc'd*, but *deeply and surprizingly affected* with the Consequences of this Nature; I was satisfy'd that they were evident *Demonstrations of Natural*, and noble *Attestations to Revealed Religion*. In which Principles the farther Improvements I have still made, or succeeding Discoveries of others have still presented to me, the more sure and certain have those Principles appeared; and the more sure and certain have those Consequences seem'd; tho' it must be confess'd that the Deepness of the Surprize and Impression, as in all the like Cases, can never be so sensible and affecting, as it was upon the first Knowledge of such amazing Truths, and momentous Corollaries from them. And I cannot but heartily wish, for the common Good of all the *Scepticks* and Unbelievers of this Age, that I could imprint in their Minds all that real Evidence for Natural and for Reveal'd Religion that now is, or during my past Enquiries has been upon my own Mind thereto relating: And that their Temper of Mind were such as that this Evidence might afford them as great Satisfaction as it has my self. For then I am sure they would not wonder at my warm and zealous Endeavours, even at the Hazard of all I have in this World, for the Restoration of true Religion, for the rescuing the Wicked out of their dangerous State, and for the bringing as many as possible to that future Happiness; which is the grand Design of Religion, and the ultimate Felicity of Mankind. But tho' this entire Communication of the Evidence that is, or has been in my own Mind, for the Certainty of Natural Religion, and of the *Jewish* and Christian Institutions, be in its own Nature impossible; yet I hope I may have leave here to address my self to all, especially to the *Scepticks* and Unbelievers of our Age; to do what I am able for them in this momentous Concern; and to lay before them, as briefly and seriously as I can, a considerable Number of those Arguments which have the greatest Weight with me, as to the hardest Part

of

of what is here defired and expected from them;
I mean the Belief of Reveal'd Religion, or of
the *Jewish* and Chriftian Inftitutions, as contain'd in the Books of the Old and New Teftament; or in all the Genuine Records now extant of both Religions. I have already obferv'd
that the *Scepticks* and Unbelievers of the beft
Senfe do now confefs, that the Arguments for a
God, his Attributes, and Providence, are very
ftrong; but they will by no Means allow, that
thofe for Divine Revelation, and for the Bible
are fo: Tho' indeed very few of them, I fpeak
it upon certain Experience, have ftudyed thofe
Sacred Books with any Degree of that Impartiality, Serioufnefs, Application, Sagacity or Piety,
which is but requifite for forming any tolerable
Judgment about them. But to wave farther
Preliminaries, fome of the principal Reafons
which make me believe the *Jewish* and *Chriftian*
Revelations to be true, are thefe following.

I. The Reveal'd Religion of the *Jews* and
Chriftians lays the Law of Nature for its Foundation; and all along fupports and affifts Natural Religion; as every true Revelation ought
to do.

II. Aftronomy, and the reft of our certain
Mathematick Sciences, do confirm the Accounts
of Scripture; fo far as they are concern'd.

III. The ancienteft and beft Hiftorical Accounts now known, do, generally fpeaking, confirm the Accounts of Scripture; fo far as they
are concern'd.

IV. The more Learning has increas'd, the
more certain in general do the Scripture Accounts appear, and its difficult Places are more
clear'd thereby.

V. There are, or have been generally, standing Memorials preserv'd of the certain Truths of the principal Historical Facts which were constant Evidences for the Certainty of them.

VI. Neither the Mosaical Law, nor the Christian Religion, could possibly have been receiv'd and established without such Miracles as the Sacred History contains.

VII. Altho' the *Jews* all along Hated and Persecuted the Prophets of God; yet were they forced to believe they were true Prophets, and their Writings of Divine Inspiration.

VIII. The Ancient and Present State of the *Jewish* Nation are strong Arguments for the Truth of their Law, and of the Scripture Prophecies relating to them.

IX. The Ancient and Present States of the Christian Church are also strong Arguments for the Truth of the Gospel, and of the Scripture Prophecies relating thereto.

X. The Miracles whereon the *Jewish* and *Christian* Religion are founded, were of old owned to be true by their very Enemies.

XI. The Sacred Writers, who liv'd in Times and Places so remote from one another, do yet all carry on One and the same grand Design, *viz.* that of the Salvation of Mankind, by the Worship of, and Obedience to the One true God, in and through the King *Messiah*: which without a Divine Conduct could never have been done.

XII. The principal Doctrines of the *Jewish* and *Christian* Religion are agreeable to the ancientest Traditions of all other Nations.

XIII. The Difficulties relating to this Religion are not such as affect the Truth of the Facts, but

but the Conduct of Providence : the Reasons of which the Sacred Writers never pretended fully to know, or to reveal to Mankind.

XIV. Natural Religion, which is yet so certain in it self, is not without such Difficulties as to the Conduct of Providence, as are objected to Revelation.

XV. The Sacred History has the greatest Marks of Truth, Honesty, and Impartiality of all other Histories whatsoever; and withal has none of the known Marks of Knavery and Imposture.

XVI. The Predictions of Scripture have been still fulfilled in the several Ages of the World whereto they belong.

XVII. No opposite Systems of the Universe, or Schemes of Divine Revelation, have any tolerable Pretences to be true, but those of the *Jews* and *Christians*.

These are the plain and obvious Arguments which persuade me of the Truth of the *Jewish* and *Christian* Revelations : which I shall briefly insist on here, and earnestly recommend them to the farther Consideration of the inquisitive Reader: Hoping that He will first endeavour to attain that Serious, Upright, Impartial, Honest, and Obedient Temper of Mind, that a Creature ought to have when he is enquiring into the Laws and Will of his Creator; and will join with me in putting up to that great Creator, some such humble Address as this following, for the good Success of his Enquiries.

" O God of my Fathers, and Lord of Mercy; Wisd. ix.
" who hast made all things with thy Word : Give 1,4,5,10.
" me Wisdom that sitteth by thy Throne; and
" reject me not from among thy Children. For
" I thy

"I thy Servant, and Son of thine Handmaid, am a
"feeble Person, and of a short time: O send her
"out of thy Holy Heavens, and from the Throne
"of thy Glory! That being present she may la-
"bour with me; that I may know what is plea-
"sing unto thee: *Amen*!

(I.) The first Reason, why I believe the *Jewish* and *Christian* Revelations to be true, is this; That they lay the Law of Nature for their Foundation; and all along support and assist Natural Religion; as every true Revelation ought to do.

That all Divine Revelation supposes the Being and Attributes of God, which are discoverable by the Light of Nature; and particularly the Perfections of Unity, Justice, Veracity, Holiness, and Goodness, all considering Men will readily grant: And that no pretended Revelation, which clearly and evidently contradicts the Laws of Nature, founded on those Divine Perfections, can be more than pretended, they will as readily grant also. So that I shall not need to prove that Part of my Proposition. But then, that the *Jewish* and *Christian* Revelations do, for the main at least, most plainly and clearly agree with, and support those Natural Notions we have of God and of Religion, and is founded upon them, is every where supposed and affirm'd in all the Original Records of those Religions. And if the Ceremonial Burdensome Laws once given the *Jews*, be objected against, as unworthy of God, and opposite to his Wisdom and Goodness, I shall take leave to say, that this is objected without any just Foundation, and contrary to the best and most authentick Accounts we have of the Reason of those Laws;

I mean

I mean, that they were given on purpose for the Support of Natural Religion; and that this appears not only by occasional Passages in the Sacred and Acknowledged Books of the Old and New Testament, but by an entire particular and noble Discourse, which we have upon this whole Subject in the *Apostolical Constitutions*, and which without all dispute is of much greater Authority than the uncertain Guesses of the Moderns. I have set down no small Parts of this Passage elsewhere, in Vindication of those Constitutions: yet are they of such Importance, that I shall take leave to repeat them in this Place.

St. Clement and Iren. Vind. or the Constitut. P. 8--11.

'We recommend to you, say the Apostles, *Ti-* '*tus*, and *Luke*, and *Jason*, and *Lucius*, and *Sosi-* '*pater*. By whom also we exhort you in the Lord, 'to abstain from your old Conversation, vain 'Bonds, Separations, Observances, Distinction of 'Meats, and daily Washings: for *Old Things are* '*passed away, behold all things are become New*.

Constitut. l. vi. c. 18.

2 Cor. v. 17.

For since we have known God through Jesus Christ, and all his Dispensation, as it has been from the Beginning, that he gave a plain Law to assist the Law of Nature; such an one as is pure, saving, and holy; in which his own Name was inscrib'd; perfect, which is never to fail; being compleat in Ten Commands, unspotted, converting Souls; which when the *Hebrews* forgot, he put them in mind of it by the Prophet *Malachi*, saying, *Remember ye the Law of Moses, the Man of God, who gave you in charge Commandments and Ordinances*; - - - -

c. 19.

iv. 4.

Now the Law is the Decalogue, which the Lord promulgated to them with an audible Voice, before the People made that Calf which represented the *Egyptian Apis*. And the Law is

c. 20.

S 4 righteous

righteous; and therefore it is called the Law, because Judgments are thence made according to the Law of Nature. - - - - This Law is Good, Holy, and such as lays no Compulsion in things Positive; for He says, *If thou wilt make me an Altar, thou shalt make it of Earth.* It does not say, *Make one,* but, *If thou wilt make*: It does not impose a *Necessity,* but gives leave for their own free Liberty; For God does not stand in need of Sacrifices, being by Nature above all Want. But knowing that as of old *Abel,* beloved of God, and *Noah,* and *Abraham,* and those that succeeded, without being requir'd, but only mov'd of themselves, by the Law of Nature, did offer Sacrifices to God, out of a grateful Mind, so he did now permit the *Hebrews*; not commanding, but, if they had a mind, permitting them; and, if they offer'd from a right Intention, shewing himself pleas'd with their Sacrifices. Therefore he says, *If thou desirest to offer, do not offer to me as one that stands in need of it; for I stand in need of nothing: for the World is mine, and the Fulness thereof.* But, after the Sin of the Golden Calf, then was God angry, as being ungratefully treated by them; and bound them with Bonds which could not be loosed; with a mortifying Burden, and a hard Collar, *&c.* - - - - that being press'd and gall'd by thy Collar, thou may'st depart from the Error of Polytheism; and laying aside that, *These are thy Gods, O* Israel; may'st be mindful of that, *Hear, O* Israel, *the Lord our God is one Lord*: and may'st run back again to that Law which is inserted by me in the Nature of all Men; That there is only One God, in Heaven, and on Earth;

and

of RELIGION.

and to love him with all thy Heart, and all thy Might, and all thy Mind, and to fear none but him, &c.

But, *Bleſſed are your Eyes, for they ſee, and your Ears, for they hear:* Yours, I ſay, who have believ'd in the One God, not by Neceſſity, but by a ſound Underſtanding, in obedience to Him that called you; for you are releaſed from the Bonds, and freed from the Servitude —— C. 21. Matth. xiii. 16.

You therefore are *Bleſſed*, who are delivered from the Curſe. For Chriſt, the Son of God, by his coming, has confirm'd and compleated the Law; but has taken away the additional Precepts; although not all of them, yet at leaſt the moſt grievous ones: Having confirm'd the former, and aboliſh'd the latter; and has again ſet the Free Will of Men at Liberty: —— And beſides, before his coming he refus'd the Sacrifices of the People, while they frequently offer'd them when they ſinned againſt him, and thought he was to be appeaſed with Sacrifices, but not by Repentance, &c. —— [Here follow many excellent Quotations to this Purpoſe out of the Old Teſtament; and then the Conſtitutions go on:] If therefore before his coming, he ſought for a *clean Heart and a contrite Spirit, more than Sacrifices*, &c. —— Not taking away the Law of Nature, but abrogating thoſe additional Laws, &c. —— C. 22.

Pſal. l. 12, 19.

For he did not take away the Law of Nature, but confirm'd it. For he that ſaid in the Law, *The Lord thy God is one Lord*; the ſame ſays in the Goſpel, *That they might know thee the only true God.* And he that ſaid, *Thou ſhalt love thy Neighbour as thy ſelf;* ſays in C. 23.

Deut. vi. 4.
John xvii. 3.
Lev. xix. 18.

in the Gospel, renewing the same Precept, *A new Commandment I give unto you, that ye love one another.* He who then forbad *Murder*, does now forbid *causeless Anger*. He that forbad *Adultery*, does now forbid *all unlawful Lusts*. He that forbad *Stealing*, now pronounces him most happy who supplies those that are in want out of his *own Labours*. He that forbad *Hatred*, now pronounces him blessed that *loves his Enemies*. He that forbad *Revenge*, now commands *Long-suffering*; not as if *just Revenge* were an unrighteous Thing, but because *Long-suffering* is more excellent. Nor did he make Laws to root out our natural Passions, but only to forbid the Excess of them, *&c.* ——— He has in several ways changed Baptism, Sacrifice, the Priesthood, and the Divine Service, which was confin'd to one Place: For instead of *daily Baptisms*, he has given only *one*, which is that *into his Death*. Instead of *one Tribe*, he has appointed that out of every Nation, the *Best* should be ordained for the Priesthood; and that not their *Bodies* should be examin'd for Blemishes, but their *Religion and Lives*. Instead of a *bloody Sacrifice*, he has appointed that *reasonable and unbloody mystical one*, of his Body and Blood, which is performed, to represent the Death of the Lord by Symbols, *&c.*

Let us therefore follow Christ, that we may inherit his Blessings. Let us walk after the Law, and the Prophets, by the Gospel, *&c.* ———

Let us be obedient to Christ, as to our King; as having Authority to change the several Constitutions; and having, as a Legislator,

of RELIGION.

giſlator, Wiſdom to make new Conſtitutions, in different Circumſtances: Yet ſo that every where the Laws of Nature be immutably preſerv'd.

(II.) The next Reaſon why I believe the Truth of the *Jewiſh* and Chriſtian Revelations, is this; That Aſtronomy, and the reſt of our certain Mathematick Sciences, do confirm the Accounts of Scripture; ſo far as they are concern'd.

'Tis certain that in many Points we can examine the Aſſertions of Ancient Authors, whether they ſay true or not, by ſome Parts of the Mathematicks; and particularly in ſuch Caſes as the Enquiries of our Age enable us to examine Things more nicely than the Authors of Old Accounts could imagine: And againſt which Methods of Examination they could therefore by no means provide; any other ways, I mean, than by giving us the Ancient Facts as they really happen'd. In which Caſes the Diſcovery of the Juſtneſs of the Proportions of Things, and the Agreement of the Old Narrations with Mathematical Computations, will be a very great Argument for the Veracity of the Writers; as will the Abſurdity of ſuch Proportions, and the Diſagreement of Things, be a like great Argument for their Careleſneſs or Falſity. Now this Reaſoning being obvious; let us try ſome of the Sacred Narrations by it; and ſee on which Side this κριτηειον will incline us, and that in ſome of its moſt remarkable Inſtances. Thus we have already ſeen that the Age of the World, as taken from the Sacred Records, perfectly agrees with the beſt Methods which Chronology and Natural

Part vii. §. 3. prius.

Natural History can afford us for its Determination.

<small>*New Theory, 2d Edit. p.144,&c.*</small>
Thus it has been elsewhere demonstrated that the most Ancient Year of the World, even before, as well as after the Deluge, had just Twelve Months, of Thirty Days apiece, or Three Hundred and Sixty Days in the whole; as *Moses*'s Account of the Deluge does most naturally imply.

<small>*Gen. vii. 11. with v. 24. and viii. 4.*</small>
<small>*New Theory, 2d Edit. passim.*</small>
Thus I have elsewhere largely shewn That that Deluge must by all Astronomical Computations, have begun that very Month; nay, that very Week, and that very Day, which *Moses* assures us it really did begin.

<small>*Chronol. Old Test. p. 12, 13, 14. and p. 55, —60.*</small>
Thus also I have elsewhere shew'd how exactly the Canon of *Ptolemy*, the surest Monument of Ancient Prophane Chronology now in the World, does agree to, and support the Sacred Chronology; and indeed illustrate the Sacred Prophecies of *Daniel* thereon depending; and that in such difficult Branches of it,

<small>*Ibid p. 198, 199, 200.*</small>
as had been otherwise too hard for all our *Jewish* and *Christian* Commentators. Many other Instances of this Nature may be also collected by the Inquisitive from my *New Theory of the Earth*; from my *Chronology of the Old Testament, and Harmony of the Four Evangelists*; and from my *Essay on the Revelation of St. John*; to say nothing of my other Writings. But because those Examples have been there already produc'd by me, I shall say no more of them here; and rather alledge a remarkable one, which has not been there mentioned.

of RELIGION.

It is well known, that *Moses* assures us, how at the general Deluge, all Land Animals that escap'd the same, were saved by an Ark; that this Ark held Sevens of Clean, and Pairs of Unclean Beasts; with their Food for the full Space of a Year; and that from thence therefore all such Creatures are now derived throughout the Earth. He also gives us the Dimensions of this Ark, 300 Cubits long; 50 Cubits broad; and 30 Cubits high. Yet does it no way appear that *Moses*, or any of his Contemporaries, could then examine the Number of all such Animals, or the quantity of Food necessary for them, during so long an Interval, as the Modern Mathematicians and Naturalists have done. Here therefore we have one of the fairest and most exact Methods of trying the Verity of this Part of the Sacred History that could well be desired: And this Method has accordingly been put in Practice in this very Case, by Two Eminent Mathematicians and Naturalists, *Buteo*, and Bishop *Wilkins*; and it has thence appeared, that if we Measure all by a common Cubit of 18 Inches, the Ark will fully contain all those Creatures, with their Food for a Year; and that in separate and convenient Cells and Apartments; though in a somewhat strait and difficult Manner. It has also since appeared, by Bishop *Cumberland*'s more exact stating of the Old Cubit, that it was not so short as 18, but in Reality near 22 Inches long; and by Consequence, as that Learned Prelate judiciously observes, that an Ark built by that Cubit, being almost double to one built by this other, will not only in a strait Manner, but

Gen. vi. vii. viii.

See Pool's *Synopsis, and Bishop* Wilkins.

Scripture Weights and Measures,

but with great Ease and Freedom, contain all that we have above-mentioned; though this it will do still, without abundance of waste Room, which would have served to no useful Purposes. So that the Result of this Computation is plainly this; That the Dimensions of *Noah*'s Ark, as set down in the Sacred History, are fully attested to by Geometry and Natural History; and that those Dimensions are as well proportion'd to the End for which they were design'd, as any Mathematician or Architect could now chuse for the like Purposes: Which Exactness, since none in the Days of *Moses* could naturally attain to, 'tis most reasonable to suppose it true Fact, and to ascribe the Direction of the whole, as *Moses* does, to the most perfect *Geometrician*, who acts always *in Number, Weight, and Measure*, the Great Creator and Provider for Mankind, and all Creatures whatsoever.

(III.) Another Reason why I believe the Truth of the *Jewish* and *Christian* Revelations, is this; That the other Ancientest and best Historical Accounts now known, do, generally speaking, confirm the Accounts of Scripture, so far as they are concern'd. That this is a most natural and sure way of discovering the Veracity of any pretended Ancient Histories in one Country, to compare them with the other Ancient and Approved Records of the Neighbouring Nations, which had Concerns with them; or with any other Ancient Records that are of good Esteem relating to the same Times, and the same Affairs, is without Controversy among all Men. Accordingly, when we are examin-

examining the Veracity of the Sacred Writers of the *Jewish* Nation, by whom almost all the Books of Scripture were written, it cannot but be highly necessary to proceed by the same Method. Nor are our Unbelievers unappriz'd of the Fitness of this Procedure, when they appear so ready to alledge even the most poorly attested Antiquities of some other Nations, to oppose them to these well attested Antiquities of the *Jews*: Which Procedure shews at once their Acknowledgment of the Reasonableness of this way of Enquiry; and yet the Weakness of their Cause upon such an Examination. For, give me leave to say, that if all the smaller Fragments, of even any tolerable Credit or Antiquity, which can be found in all the old Books, of which we have any Accounts, and may seem considerably to contradict the Sacred Records, were gathered together, and were compar'd with those most Ancient, Authentick, and Numerous Books and Fragments, which evidently support them; those I mean, collected by *Josephus* against *Apion*; by *Eusebius* in his Evangelical Preparation and Demonstration; by *Huetius* in his Evangelical Demonstration; by *Grotius* in his Truth of the Christian Religion; by *Bochart* in his *Phaleg.* and *Hierozoicon*; by Bishop *Stillingfleet* in his *Origines Sacræ*; and by many others who have written on that Noble Subject, the Difference of the Evidence would appear vastly great and surprizing. Insomuch that one of the greatest Masters of all that Ancient Learning, *Grotius*, does directly profess, that if we do not reckon some open Enemies to the *Jewish* and *Christian* Religions who,

De veritat. Relig. Christ. iii. 14.

who lived too late to know Things themselves, and were too partial to be believ'd without other Authority, there are not any Genuine Records, or Testimonies of Antiquity extant, that contradict the Scriptures. Which Thing being so, it is most highly reasonable for us to have a great Veneration for those Sacred Records, which, however different from Modern Histories, as they ought to be, are yet so very agreeable to, and so fully confirm'd by the other oldest and most faithful Remains of the Ancient Ages of the World.

IV. Another Reason why I believe the Truth of the *Jewish* and *Christian* Revelations is this; that the more Learning has increased, the more certain in general do the Scripture Accounts appear, and its difficult Places have been more clear'd thereby. If Hypotheses, or Accounts of Things in any Kind, especially when they are strange and surprizing, be advanc'd or believ'd by any, and then put into the Method of Examination before the World, when it is improving in Knowledge, and New certain Discoveries are frequently made in all Parts of Learning; it is justly to be expected, that what is really true and well-grounded will stand the Test; and what is not so, will sink under it: The one will thereupon be approv'd and established; and the other rejected and discarded. Thus it happens frequently in Human Opinions; and by this means the Improvement of the Learning of these Two last Centuries has done vast Service to Truth, by distinguishing what is Solid and Genuine; from what is Trifling and Spurious. Thus we we now generally know which Writings pretending to Antiquity, or to belong to Ancient Authors of Reputation, are Genuine, and which are Suppofititious; which till the late Revival of Critical Learning, were ftrangely confounded together. Thus the late Improvements in Aftronomy, particularly by Telefcopes, and the wonderful Difcoveries of Sir *Ifaac Newton*, have determin'd the Difpute between the *Ptolemaick*, the *Tychonick*, and the *Copernican* Syftems of the Heavens, in Favour of the laft; which till lately was Matter of great Difpute, even among the Aftronomers themfelves. And a great Number of other Examples might eafily be alledg'd to the fame Purpofe. Let us then apply this Method of Trial, and fee whether the Ancient Polite Learning of the *Greeks* and *Romans* of Old; or the much greater and folider Learning of the Two laft Centuries; the former of which grew up fomewhat after the Times of the Old Teftament, and the latter long after the Time of the New, has confirm'd or contradicted thofe Sacred Books of the *Jews* and *Chriftians*; that we may the better judge of their Solidity and Authority. Now in this Enquiry, what *Jofephus* has produc'd for the *Jewifh* Religion, in his Books againft *Apion*, already quoted, may ferve as a Specimen of the vaft Advantage the *Jewifh* Infpired Writings received from the *Greek* and *Roman* Learning; of which he was a great Mafter. And what *Grotius* has produc'd for both the *Jewifh* and *Chriftian* Religion, and the clearing their Difficulties in his *Truth of the Chriftian Religion*, and Comments on the Scripture, may be a Sample of the great Confirmation they all receive from the Revival of Ancient Learning in the last Ages; of which he was no less a Master. Nor, as I hope, will what has been above produc'd from the Modern Astronomy, Mathematicks, and Philosophy, for the Support of the same Inspired Writings, be unfit to be esteemed a farther Specimen in general to the same Purpose. But that I may not seem wholly to content my self with what has been already observed by my self or others, under this Head, I shall produce a new Specimen or Two for the same Purpose. 'Tis well known what a Noise the Scepticks and Unbelievers make with the Uncertainty of very Old Accounts, and very Old Books; the great Omissions, Additions or Interpolations that may easily have happen'd in them, during a Course of many Ages; and the consequent Uncertainty of the Books of the New, but especially of those of the Old Testament; and so most of all of the oldest Parts of it, such as the Five Books of *Moses*. Now in this Case the Revival of Learning, and of Languages, and the Inquisitiveness of this last Age, has procur'd us a most Noble Treasure, as an Attestation in this Matter, where it was least expected: And from the small Remainders of the Old *Samaritans*, still left about *Sichem* in *Judea*, we have lately recover'd the *Pentateuch* it self: Not as in later Ages, known and owned by the Body of the *Jewish* Nation, or Two Tribes, spread over the World; but as peculiarly receiv'd and allowed by all the Twelve Tribes, before the *Babylonish* Captivity;

as

as written in the Original Character used before that Captivity, by that whole Nation; and as continued down to, and received by the Remains of those Ten Tribes in *Samaria*, all the Times of their bitter Hatred to the other *Jews*, even to this very Day. So that this must needs be a Copy entirely distinct from the common *Massorite* one now current, and in all Probability much older than those from whence even the *Septuagint* made their Translation, long before the Times of our Saviour. Yet upon the Comparison of this *Samaritan Pentateuch*, with the common *Hebrew*, and with the *Septuagint*; abating the Diversity of some Chronological Numbers in them all, as to the Lives of the Patriarchs, before and after the Flood, of small Importance here, there appears but very little Difference between them; and none at all that I know of in any Points of Consequence, either in History, Doctrine or Practice. So that, for the main, it exceedingly strengthens the Authority and Uncorruptness of our present Copies of the *Pentateuch*; and by Parity of Reason, of the present Copies of all the Books of the Old Testament, which do with us otherwise stand upon the same Foot, and are usually of a much lower Antiquity than these Five Books of *Moses*. 'Tis also well known that the Scriptures affirm, what seems to us very surprizing, and next to impossible, *viz.* that in the early Times of the World, Men commonly liv'd to many Hundred Years of Age; and that in particular before the Deluge, they liv'd frequently to near a Thousand; that after the Deluge, their Lives declin'd gradually from

Gen. v. *and* xi.

Five or Six Hundred, to about Seventy or Eighty Years only; although from the Days of *David*, till this Time, that Age of Man has been at a Stand, and about that smallest Duration. This Account must be own'd to be bold and strange to us, after near 3000 Years Experience of the last and shortest Period of Human Life. But then, when we reflect that the Learning of *Josephus* produc'd many strong Ancient Attestations to this Account, which were then Extant; that the present *Chinese Annals*, lately discover'd, exactly agree, so far as they are concern'd; as I have elsewhere shewn; and that the present Numbers of Mankind upon the Earth, taken together with the usual Period of Years for their Increase and Doubling; and with the Number of Years that by the best Evidence the Earth has been in its present State; do plainly require such longer Duration of Mens Lives in those Ancient Ages: Which last ways of Trial are entirely owing to the new Improvements in Learning: We shall see Reason at once to believe the Truth of the Fact, how strange soever it may now appear; and to pay a due Deference to those Original Sacred Records, whence we were first and best inform'd of it. It cannot also but appear very strange at first Sight in the Old Testament, that *David* and *Solomon* especially, should in so peculiar a Manner possess and make Use of a much greater Quantity of Silver and Gold than we have any like Examples of even in this Age; when yet by our Modern Navigation, the *Indian* Mines have yielded us such new Treasures of that Kind; and that yet afterward

Antiq. I. 4.

Chronol. p. 60--65.

p. 65--68.

ward the *Jewish* Nation should lose those mighty Riches. Yet will all this appear very agreeable to Truth, if we consult Dr. *Prideaux*'s late Admirable Book, *Of the Conjoining the History of the Old and New Testament*, where in the very Beginning of it, he has trac'd the *Jewish* Navigation, and such its Effects, with the greatest Sagacity, and to the greatest Satisfaction.

I might also instance in the Improvements in Sacred Chronology, by the Discovery of the Canon of *Ptolemy*; in the full Solutions of the Deluge long since past, and of the Conflagration yet to come; with many other Things of a like Nature, which the Modern Improvements and Discoveries have assisted us in; and all still in exact Agreement with the Scriptures of the Old and New Testament: But because I have already hinted at these Matters under former Heads, and have elsewhere more fully insisted on them to the same Purpose, I shall no farther enlarge on them in this Place.

(V.) I believe the *Jewish* and *Christian* Revelations to be true, because there have been generally such standing Memorials preserv'd of the Truth of the Principal Facts, as give us great Assurances they were real. That this is a proper and usual way of preserving the Memory of past Actions, the Customs, and Medals, and Pillars, and Inscriptions, and Solemnities, and Sepulchral Monuments of all Nations, do Testify. And that both the *Jewish* and *Christian* Legislators have remarkably taken the like Care and Method, I have elsewhere observ'd;

Primitive Christ. Reviv'd, Vol. iii. C. 2. §. 4. p. 174--179.

obferv'd; to which I refer my Reader. Only give me leave to Inftance here, in a few of thefe ftanding Memorials or Monuments which are not there taken Notice of, and which feem to me remarkable Confirmations of the Truth of the Sacred Hiftory, even in fome of its leaft probable Branches. Thus the Accounts we have in *Genefis* iii. of the Fall of *Adam*, upon the Temptation of the Devil under the Appearance of a Serpent, and the Suggeftion of his Wife; and the confequent Change of the State of our firft Parents thereupon, is, in all its Branches, one of the moft improbable and amazing Hiftories in the whole Bible. I mean even as taken barely and literally, and as expounded in the Old Chriftian Records themfelves; without thofe abfurd Additions and Improvements which *Auftin* and *Calvin*, with their Followers, have joined to it. Yet when I confider that the Remains and Memorials of this great Cataftrophe already mentioned, are evidently true in Fact, and yet can no way be accounted for on any other Hypothefis; I fubmit my Faith to the Evidence of the Sacred Hiftory, fo well attefted to by the prefent State of Nature; and rather wait with Patience till Providence fhall pleafe to unravel the Myftery of this furprizing Scene, than venture, by denying the Truth of the Facts, to oppofe my felf to that ftrong Evidence, which the *Mofaick* Hiftory, as attefted to by thofe fenfible Demonftrations, affords us. Thus the *Dead Sea*, or Lake of *Sodom*, with its known, but unexampled and furprizing Phænomena, look as if it were the direct Remains of a terrible Earthquake, join'd

with

Part vii. §.4. p.145. prius.

See Bifhop Patrick on Gen. xix. 24, 25.

with a more terrible Shower of Fire and Brimstone from Heaven, by which *a fruitful Land was turned into a Salt and Sulphureous Sea, for the unnatural Wickedness of them that dwelt therein:* Psal. cvii. 34. Exactly according to the Sacred History of that terrible Destruction of those People, who at once Inhabited and Polluted that Noble Soil; and is become a seasonable and standing *Example, having suffered the Vengeance of an eternal or unquenchable Fire:* Jude v. 7. Or, in the Words of the Author of the Book of Wisdom, *Of whose Wickedness even to this Day, the wast Land that smoaketh is a Testimony; and Plants bearing Fruit, which never come to Ripeness; and a standing Pillar of Salt, a Monument of an unbelieving Soul.* x. 7.

Thus the Annual Feasts of the Passover, of Weeks, and of Tabernacles, among the Ancient *Jews*, together with the Annual Baptism, and Weekly Communion among the Ancient Christians; with their Solemnities of *Easter, Ascension,* and *Pentecost*; besides their immediate Application to the Uses of Piety and Religion, did also admirably and constantly attest to the Truth of those wonderful Facts, whose Memorials they were; I mean the miraculous Deliverance of the *Jews* from the *Egyptian* Bondage, with their Legislation, and living in Tabernacles for Forty Years in the Wilderness afterwards; and the Death, and Resurrection, and Ascension of our Saviour; on which Facts the Two Institutions are principally founded. I might be very large and full under this Particular: But that would make this Argument too Disproportionate to the rest, and would lead me too far out of my Way.: So I forbear; and proceed to Declare,

(VI.) That I therefore believe the *Jewish* and *Christian* Revelations to be true, because I am persuaded that neither of those Religions could possibly have been receiv'd, establish'd and preserv'd in the World, without such Wonders and Miracles as the Sacred History contains, and which are undeniable Proofs of their coming from God. For, to say nothing here of the *Gentiles*, if we Consider but the Obstinacy and Perverseness of the People of the *Jews* in all Ages, and that still they did, to the utmost Times we can trace them, in general, and as a Nation, most firmly believe, and openly submit, and most tenaciously adhere to the Doctrines, Discipline, Laws and Government delivered by *Moses*; how strange, uneasy, or burdensome soever, several of the Particulars were; and did yet in great Numbers, on the Preaching of the Gospel, renounce many of those their former Opinions and Constitutions, of which otherwise they were so exceeding tenacious; and entirely receiv'd, believ'd and obey'd the Christian Revelation, on its first Appearance in the World; and this in Opposition to their former Prejudices, their certain Interests, their Reputation, and Desire of Self-preservation; nay at a Time when they knew they must frequently lose, and suffer, and die for this new Religion; and all this without any other Hopes than what belong'd to another World; and must be entirely frustrated, if that Religion proved false: Considering all this, I say, it is, morally speaking,

ing, impossible that such Numbers, of *Jews* especially, should submit successively to both these Religions, as they certainly did, without convincing and undoubted Miracles for their Confirmation.

(VII.) I believe the Truth of the *Jewish* Revelation, because I perceive that that Nation, which all along Hated and Persecuted the Messengers and Prophets of God when they were alive, were yet forced to believe all along that they were true Messengers and Prophets of God, and their Writings of Divine Inspiration. This is to me a very remarkable Observation, a very Certain one, and of the greatest Consequence in this Enquiry. That the *Jews* were anciently a Stubborn, Disobedient, and Stiff-necked People, and not without the most forcible Methods to be reduc'd to the Observance of the Divine Laws, is evident in their whole History; that they all along, in particular, opposed *Moses*, and Rejected, and even Persecuted and Slew, the succeeding Prophets, when they were sent unto them to call them to Repentance and Amendment, is alike evident therein: That yet they were forced to own, that these Men were the true Prophets of God, is equally plain therein; and is demonstrable from their constant Reception afterwards of those Writings wherein these Things appear, as Divinely Inspir'd, to this very Day. It is also clear in those very Writings still extant, that their Contents are so cutting and severe, without the least Tincture of Flattery, or sparing them in their Vices, that nothing but a full Conviction of the Certainty of their Divine Mission and Authority, could ever
induce

induce them to a Reception of them. The Case in general of the *Jewish* Legislator and Prophets, was all along much the same with that of good *Micaiah*, in the Days of *Ahab* ; concerning whom when *Jehosaphat* enquired, *Is there not a Prophet of the Lord that we may enquire of him?* Wicked *Ahab* reply'd, *There is One Man*, Micaiah, *the Son of* Imlah, *but I hate him; for he doth not prophesy Good concerning me, but Evil.* Yet all this Hatred notwithstanding, *Micaiah*'s Denunciation of God's Judgments upon *Ahab*, followed with its immediate and dreadful Completion, soon convinc'd them all, that he was a true Prophet of God: And He is accordingly ever since allow'd to have been such by that whole Nation. And this has been equally true of the rest of the *Jewish* Prophets all along: which seems to me a strong Argument for their Veracity and Authority, as still absolutely undeniable among that Nation.

1 Kings xxii. 7, 8.

v. 19--38.

(VIII.) I believe the *Jewish* Religion to be True, because the Ancient and Present State of the *Jewish* Nation shews the Truth of their Law, and of the Scripture Prophecies relating to them. The Law of *Moses* did, in the plainest and most affecting manner, deliver such obliging Promises to that Nation, upon their keeping close to the Worship of the True God, and to the Obedience of the Laws then delivered to them; with such terrible Threatnings upon their Rejection of them, upon their Idolatry and Disobedience; and the succeeding History and State of that Nation all along, does so clearly and particularly inform us of the Completion of those Promises and Threatnings to this very Day, as afford

afford us the strongest Evidence for the Divine Authority of those Denunciations. [See Bishop *Patrick* on *Deut.* xxviii]. Nor is the Case different in the particular Prophecies occurring in all the Old Testament, relating to their Descent into, and sojourning in *Egypt*; their *Babylonish* Captivity under *Nebuchadnezzar*, and their Restoration under *Cyrus*; with their present long, and tedious Banishment from their own Country, without a Prophet, or King, or any Divine Revelation; with many other Circumstances of their Affairs, which have been evidently fulfilled in their proper Seasons. This Argument seems also to me of very great Weight. But because I have elsewhere insisted on it, in my *Boyle's Lectures*, and in my *Chronology of the Old Testament*, I shall not here enlarge, but refer the inquisitive Reader thither for his farther Satisfaction. Only I may be allow'd here to put the Unbelievers in mind, how peculiar and unparallel'd the State of this People has ever been. That at first they were separated from the rest of the World; that they lived in it afterward like a separate Species of Creatures; that after all their Miseries and Captivities, they have still preserved themselves a separate People, and still recovered their ancient Land and Settlement again, in a wonderful manner, till this their last grand Captivity and Dispersion; and that the Body of the two Tribes, and some Remains of the Ten at least, are even now, to the Surprize of all considering Men, after some Thousands of Years, a separate Body, unmixt with other Nations, among which they sojourn; as at once a standing Monument of the Truth of their original

Sacred

Sacred Books, and waiting in readiness for their final Restoration, according to the same Prophecies.

(IX.) I believe the Christian Religion in particular to be true, because the Ancient and Present State of the Christian Church shews the Truth of the Original Records of Christianity, and of the Scripture Prophecies relating to them. The Author of our Religion, and his Apostles, referr'd the *Jews* to the many Predictions concerning their *Messias* occurring in the Old Testament; particularly, that famous one now known by the Name of *Daniel*'s 70 *Weeks*, which they could not then deny to be fulfilled in him: They foretold that Opposition, and Persecution, which the Christian Religion should meet with in its Propagation; that yet it should gradually prevail over the World, notwithstanding such Opposition and Persecution: That the Body of the *Jewish* Nation, with their Temple and Worship, should, for a Punishment of their heinous Sins in Crucifying their *Messias*, and Persecuting his Followers, in that very Generation of Men, be utterly subverted and destroyed; that the *Jews should fall by the Edge of the Sword*; *that they should be led away Captive into all Nations*; *and that* Jerusalem *should be trodden down of the* Gentiles, *till the Times of the* Gentiles *should be fulfilled*: That the Church of Christ should at first be Pure and Holy: But that after some time false Teachers and Hereticks should corrupt it: That at *Rome* or Mystical *Babylon* especially, a Grand Apostacy or Antichristian State should arise, and over-bear and corrupt true Christianity, and under Ten Antichristian Kingdoms, introduce Tyranny and Idolatry into it for 1260 Years

[margin: Mat. xxiv. 15. Mar. xiii. 14. 10. Dan. ix. 24, &c. Mat. xxiv. & passim alibi.]

[margin: Luc. xxi. 24.]

[margin: Mat. xxiv. 2 Theff. ii. 1 Joh. iv. 3. Apoc. paffim.]

Years together; till after which time the Church should not be able to recover her Primitive Purity; with many other such great, and strange, and, when the Christian Books were written, most highly improbable Events; all which have eminently and notoriously come to pass in the Face of all the World, as the most convincing Demonstrations of the Verity and Divinity of the Christian Religion. I have elsewhere largely treated of the greatest Part of these Predictions, in my Essay on the Revelation of St. *John*, and shewn the Exactness of their Completion, I shall not therefore enlarge any farther upon them in this Place.

(X.) I believe the Christian Religion to be true, because the Ancient Adversaries of Christianity acknowledg'd the Miracles whereon the Christian Religion is founded, to be True. I here insist only on the Confession of Adversaries as to the Truth of the Facts on which the *Christian Religion* depends: Not that I doubt of the like Confession as to the Miracles wrought by *Moses* and the Prophets, for the Confirmation of the *Jewish* Religion, where-ever their Adversaries had means of knowing the same. But because we have but a few Records, or rather Fragments of Records now extant from the Heathen Writers, contemporary with those of the Old Testament; and of them still fewer that relate distinctly to the ancientest and principal *Jewish* Miracles, I chuse to confine this Observation to the Christian Miracles. And that these Facts themselves were originally allow'd to be true by the Enemies of Christianity, appears, not only from the Conversion of vast Multitudes of those Enemies by the convictive Evidence of such Miracles to the Faith of Christ, which I have shewed could not have happen'd if they had not been satisfy'd of their Reality; but by the History of the New Testament, and the Mention there of the Objections made against that Religion, which all supposed the same Reality; and by the Remains of the ancientest Writers against Christianity, whether among the *Jews*, such as *Trypho*; among the Heathens, such as *Celsus*; or among the old Hereticks, such as *Simon Magus*; which do almost unanimously confess, and take for granted, that Christ and his Apostles did work such Miracles for the Confirmation of the Christian Religion, as the Christian Records do plainly testify at this Day. And I dare appeal to all our present Scepticks and Unbelievers, whether they can possibly persuade themselves, as the Old Infidels did, that the Christian Religion is false, notwithstanding there were very many real Miracles certainly wrought for its Confirmation: I believe they are not so weak; and I hope not so wicked neither as this comes to.

See Just. Dialog. with Trypho. Origen against Celsus. Recognit. passim.

(XI.) I do therefore believe the *Jewish* and *Christian* Revelations to be true, because the Sacred Writers, who liv'd in Times and Places so remote from one another, do yet all carry on One and the Same Grand Design, *viz.* that of the Salvation of Mankind, by the Worship of, and Obedience to, the One true God, in and through the King *Messiah*; which without a Divine Conduct and Inspiration, reaching through those Ages and Places, could never have been done. This Observation is very remarkable as to the *Jewish* and *Christian* Writings, if they be compar'd with

with the several Books of the *Greek* Philosophers; the former of which never dispute or debate what their main Scope was to be, or what they were to have ever in View, as ever naturally tending to one known Design beforementioned; while the latter were so far from any common Principles of that Nature, that the very Design they were to aim at, or the *Summum Bonum* it self, the supreme Happiness of Man, was a Matter of wonderful Debate among them. Nay, *Varro*, one of their most famous Authors, is said to have reckon'd up no fewer than 288 possible Opinions about it. And 'tis very plain, that the due Worship and blissful Enjoyment of the One True God, and that by the Means of the King *Messiah*; or indeed any other such Noble and Divine Means of attaining them, are almost wholly Strangers to the Heathen Philosophers. And no wonder, when they were so far from knowing the Nature and Will of the One True God, with the true Worship and Obedience due to him, which yet were knowable to Inquisitive Honest Minds by the Light of Nature; or from owning the Necessity of a Mediator, or of Divine Revelation; that they did not generally acknowledge the One True God himself; much less did they make the Worship of, and Obedience to him, the Foundation of their Doctrine; as all sound Philosophy, as well as Divinity, ought certainly to have done. Accordingly, tho' they all agreed in the Excellency of Virtue in general, which God has too deeply stamp'd upon Human Souls, and made too evidently necessary in general to Human Happiness, even in this World, to permit

Ap. August. De Civitat. Dei.L.xix. 1.

mit it to be overlook'd by any thinking Men; yet did they agree in almoſt nothing elſe. And indeed, their ſeveral Philoſophical *Dogmata*, ſeem like our Modern School-Divinity, to have been rather creditable Kinds of Amuſements, and Subjects for the Exerciſe of Wit and Parts in Diſputation, than directed to the real Inſtruction and Improvement of Mankind in true Religion, or in the Attainment of Happineſs, either in this or in another World; which is the main View of the Holy Scriptures. And accordingly the Variety of Opinions, and the Frequency of Diſputes among thoſe Philoſophers, did but nouriſh this diſputacious Humour of the ſeveral Parties; and this without any Proſpect of the real Diſcovery of Truth, or of reducing Men to a regular and religious Way of Life, in order to their future Happineſs; which for certain ought to be the grand Intention of all our Philoſophical and Religious Enquiries. Let but any one compute the unanimous Agreement of all the Sacred Writers, from *Moſes* to *Polycarp*, in the main Scheme of Divine Revelation, and Conduct of Human Life; with the almoſt entire Diſagreement and Uncertainty there is as to ſuch important Matters among the Heathen Philoſophers; ſo far I mean as thoſe Matters are known to be deriv'd from Revelation; and he will ſoon ſee a vaſt Difference between them; and will not be able to account for it, without allowing the former to be of Supernatural and *Divine*, and the latter of bare *Human Original*.

(XII.) I believe the Truth of the *Jewiſh* and *Chriſtian* Revelations, becauſe the principal Doctrines therein delivered are agreeable to the
ancienteſt

of RELIGION.

ancientest Traditions of all other Nations. For, tho', as I have just been observing, the Heathen Philosophers among the *Greeks*, who were comparatively later, and more modern, and who followed their own Reasonings in all such Matters, were mighty uncertain, and various in their Philosophical and Religious Notions; as all Men are when they have no better guidance than Human Supposal and Conjecture; Yet was it quite otherwise with the more ancient Ages, and those Natural and Divine Doctrines which they received by Tradition from their first Founders, and which most probably were originally deriv'd from the first Parents of Mankind, or at least from the earliest of their Progenitors after the Deluge. Those I mean whose Traces and Fragments are still extant in the earliest Sacred Books of the *Egyptians*, *Druids*, *Tyrians*, and *Brachmans*, in the Remains of *Trismegistus*, of *Orpheus*, and *Zoroastres*, &c. and in the *Sibylline* Oracles; those Parts, I mean, of them which are well attested to by *Heathen*, *Jewish*, and *Christian* Antiquity. These most ancient Traditions, as has been largely and fully shewn by *Grotius*, Bishop *Stillingfleet*, and many others, do, for the main, admirably agree with the *Jewish* and *Christian* Revelations; not only as to the particular Histories and Facts contained in the *Old Testament*, of which already; but as to the principal Points on which those Religions are grounded; I mean, the Unity and Attributes of God; the Creation of the World by him; its Deluge already past, and Conflagration still future; the Immortality of Human Souls; and the Judgment to come; with the Rewards and Punishments of the next World. This Agreement of the most Ancient Heathen Traditions,

Traditions, and that in several Parts of the World, with the like Contents of the Sacred Writings of the *Jews* and *Christians*, cannot but be a mighty Attestation to them, both as to those particular momentous Points themselves, and also, in a good Degree, to the rest of their Contents, so far as they are any way connected with, or belong to them. Insomuch that He who is an Infidel, in those fundamental Articles especially, must not only oppose himself to the Inspired Records of the *Jews* and *Christians*, but also to the best and oldest Remains we have in all others Nations relating to the same Doctrines.

(XIII.) I believe the Truth of the *Jewish* and *Christian* Records, notwithstanding the Difficulties thereto relating, because I observe that those Difficulties do not affect the Truth of the Facts, or Assertions, on which those Religions are grounded, but the Conduct of Providence only; the Reasons of which Conduct are no Parts of those Religions; and accordingly, the Sacred Writers do never pretend fully to know, or to reveal them to Mankind. This I take to be an Observation of great Weight, and yet not sufficiently taken notice of by any; *viz.* That those Inspired Writers, who deliver us the most important Messages and Commands in the Name of God, with the utmost Assurance; who relate the most surprizing Miracles, and that as done, or seen by themselves, and many others openly, with the greatest Boldness; who denounce Threatnings, or promise Blessings, quite beyond their own Ability to make good, with the most assured Confidence; who reprove Princes and People as to Crimes of the highest Nature,

ture, even while they were entirely under their Power, and in their Hands, with the utmost Freedom and Courage; and who in their whole Conduct ever show a perfect satisfaction in the Truth of their Mission, and Certainty of their Inspiration from God; do yet rarely or not at all meddle with the Reasons of Providence, the Justification of the Divine Orders, or the Vindication of the Justice and Goodness of God in such his Dispensations by them. Nay, the rest of the Prophets or Apostles, excepting our Blessed Saviour himself, do generally seem equally puzzled and surpriz'd at several of those Dispensations; and are found as ready to expostulate, tho' generally in a decent and most humble manner, with God, concerning such his strange and surprizing Procedure, as other Men, no way concern'd in any such Divine Dispensations at all. Thus, for Example, we find in the ancientest Book now extant in the whole Bible, and probably in the whole World, I mean the Book of *Job*, the same Difficulties and Disputes about the Conduct of Providence, the Prosperity of the Wicked, and Afflictions of the Righteous, that we every where else meet with in other Authors. And, what is most of all remarkable, we here find, that when God himself is introduc'd, as finally determining those Debates, it is done without the Assignation of the particular Reasons for this Procedure; any farther, than the noble Representation of the Power and Wisdom of the Almighty in general, and the Meanness and Inability of his Creature Man, and the consequent Submission due from the one to the other, may be thought sufficient for that Purpose. Thus we find the *Psalmist* equally sur-

Job xxxviii. xxxix. xl.

Pf. lxxiii.

priz'd

priz'd at the same strange Conduct of Providence, and hardly able to extricate himself from the Difficulties therein; even with all the Advantages of the *Mosaick* Religion, and the more constant Happiness of the Good, and Afflictions of the Bad under it, than in any other Nations of the World. Thus also we find the Prophet *Jeremiah*; one whom God frequently made use of in his Revelations to the *Jewish* Nation, and who endured the severest Imprisonment for his Faithfulness to his Duty as a Prophet; was equally shock'd and surpriz'd at the same seemingly unequal and partial Conduct of Providence; and complains to God of it; yet without any full or satisfactory Answer: And the like Observations may be made as to *Solomon*, and other of the Inspired Writers. While yet, these their Difficulties and Complaints, never in the least made them doubt or dispute about the Certainty of their Mission and Inspiration; about the Truth of the Promises and Threatnings they denounced from God; or about the Reality of the Miracles that were performed in his Name. Accordingly I observe, that *Simon Magus*, with his old Followers among the Hereticks, who allowed the Truth of the Facts and Miracles of the Gospel, did only make use of such Reasons as were taken from the seeming Injustice, and Unaccountableness of several Parts of Providence, in order to subvert Christianity. Nor is the Case much different among our present Unbelievers; who being not able to find any good Grounds to overthrow the Truth and Certainty of the Prophetick and Miraculous Attestations, which the *Jews* and *Christians* alledge for the Divinity

Margin notes:
Jer. xii. 1. &c.
Eccles. passim. See 4 *Esd.* passim.
Recogn. passim.

Divinity of their Religion, do commonly proceed after the same manner, and muster up all the Arguments they are able from the like Instances of Providence, or Passages in Scripture, which seem to them absurd and unreasonable. Now what is the natural Result of all this? But that, as the Scriptures every where allow and suppose, God's proper Time for unravelling the Mysteries of his Providence, *the Day for the Revelation of his righteous Judgment*, is not yet come; that he has yet during all this Interval, afforded sufficient Demonstrations of the Certainty of the *Jewish* and *Christian* Revelation's Derivation from him notwithstanding; which is the proper State of this Matter, and ought to be the proper Measures of our own Faith and Practice accordingly. For as the Difficulties are there and only there, where Divine Revelation does not pretend they are yet cleared; so is there no certain Difficulties, so far as the same Revelation pretends to Evidence and Demonstration, for what it recommends to us; which Case methinks highly deserves the Consideration of our modern Scepticks and Unbelievers.

Rom. ii. 5.

(XIV.) Natural Religion, which is yet so certain in it self, is not without such Difficulties, as to the Conduct of Providence, as are objected to Revelation; and therefore none that believe the former, ought to be deterr'd by such Difficulties from believing the latter. This is also a somewhat uncommon, but certainly not an improper Observation, as to the Credibility of Divine Revelation, to all those I mean who are not entirely Atheists, or against all Divine Providence. Thus we find in the Conduct of

Natural Providence some such strange *Phænomena* as are not easily accountable, or reconcileable to the most becoming Notions we all have of the One, Wise, Just, and Beneficent Creator and Governor of the Universe. We find such Antipathies of even one brute Creature against another; such a vast Number of those Creatures destroy'd immediately upon their Birth; such sudden and untimely Deaths that frequently happen to many others of them in the Course of Things; such a vast Number of them devoured by other brute Creatures, by Men, nay, sometimes by those of their own Species: We find in the State of Mankind, even abstracted from Revelation, so many Miseries and Calamities every where in the World, involving the several Individuals from their Birth to their Grave, and those often no way to be avoided or cured by any Methods of Prudence, or Virtue, or Religion it self, as affords great Difficulties to thinking Persons, and have occasion'd abundance of Hypotheses, in order to their Reconciliation with a general Providence: And after all, are in great Measure equally difficult to us, as they have been to the past Ages of the World. If therefore, these great Objections notwithstanding, the Arguments for Natural Providence do still appear cogent and undeniable, as indeed they now do more than ever; and all wise Men rather chuse to believe what they have full External Evidence for in Natural Religion, although they cannot yet solve all the Difficulties therein, why do not we proceed after the same Fair and Impartial Manner in the Business of Revelation? And equally believe those Scriptures, for which we have such strong and cogent Arguments,

although

although we cannot yet solve all the Difficulties contained in them? Especially when I may venture to say, the Case of Divine Revelation, compar'd with Natural Religion, is, as to this Point, not at all to the Disadvantage of the former; and that the Objections against Natural, are not at all Inferior to those against Reveal'd Religion: As any Impartial Man will easily confess upon the Comparison.

(XV.) I therefore believe the *Jewish* and *Christian* Revelations to be true, because the Records of the same, or the Books of the Old and New Testament, have the greatest Marks of Honesty and Impartiality of all others; and withal, have none of those known Marks of Knavery and Imposture, which all false and spurious Writings must certainly have. These Observations seem to me highly useful, and very certain and obvious, upon a careful Perusal, and exact Comparison: But because they are common, and frequently insisted on by others, and if enlarg'd upon, would take up too great a Room here, I shall chuse to refer the Reader to those who have already treated of those Subjects; particularly to our excellent Dr. *Prideaux*, in his Appendix to the Life of the Impostor *Mahomet*.

Only I beg of the Inquisitive Reader himself to reflect on this Head, as he reads the Sacred Histories, and to consider with himself how very improbable it is that those Sacred Accounts, which of all other Ancient Writings now Extant, have evidently the greatest internal Marks of Sincerity, and the least imaginable Signs of Falshood of all others, should yet be the grossest Forgeries, and most notorious Impostures in the whole World: As upon the Supposition of their being untrue, they must most certainly appear to be to all Mankind.

(XVI.) I believe the Truth of the *Jewish* and *Christian* Revelations, because the Scripture Predictions have been still fulfilled in the several Ages of the World whereto they belong. This is an eminent and open Method of trying the Truth, or Falshood, of any pretended Revelation, which is so explicit as to foretel future Events; especially those that are very remote, and depend mainly on the free Actions of Men, or on the Counsels of God, and the State of the Invisible World; such as the *Jewish* and *Christian* Revelations most certainly are. And to this Completion do I venture openly to appeal, for the Justification of those Institutions. Accordingly, I dare venture to affirm, with St. *Peter*, that this Character of a *more sure Word of Prophecy*, is one of the strongest Arguments for the Truth of the Scriptures, of all other whatsoever. I have already shew'd elsewhere, how exactly the Sacred Predictions that belong to Times already past, and were extant in the Days of *David*; as also not a few of those belonging to Christianity afterward, have been exactly fulfilled in their appointed Times, in my *Sermons at Mr. Boyle's Lectures*; in my *Chronology of the Old Testament, and Harmony of the Four Evangelists*, and in my *Essay on the Revelation of St.* John. And I declare I am so far from seeing any Reason from the present Posture of Affairs in the World, to doubt of the Completion of those which remain, even for the main, as I have expounded them, that I rather find great Cause to believe, that the Prophetick Scheme begins to clear up apace, and that the
Kingdoms

2 Pet. i. 19.

Kingdoms of this World, as is there Predicted, will in no long Time (and this probably, as moved in Part by the Plainness of the Completion of some of the Sacred Predictions just now past, or soon coming on) *become the Kingdoms of our Lord, and of his Christ; and that he shall Reign for ever and ever.* But this important Matter has been so particularly treated of by me elsewhere, in the Treatises already referr'd to, that it is by no means proper to enlarge upon it here any farther; Only, *Whoso Readeth, let him Understand. He that hath Ears to Hear, let him Hear.*

Apoc. xi. 15.

Mat. xxiv. 15. xi. 15.

(XVII.) I am therefore obliged to admit the *Jewish* and *Christian* Revelations to be true, and Divine; because no opposite System of the Universe, or Scheme of Divine Revelation, has any tolerable Pretences to be true, or can be compar'd, as to Evidence, with those of the *Jews* and *Christians*. Whither would our Atheists or Deists have us go for Information and Satisfaction, in our most concerning Enquiries about the State of the Universe, and of Religion, and of our Souls hereafter, if we must discard the Holy Scriptures? Must we go to any other Records, as better attested to? This I believe they will not say. Must we have Recourse to the exploded and absurd Schemes of the Eternity of the World, and the Fatality of all Things; or to that greater Absurdity of the Formation and Dissolution of the Universe by the Fortuitous Concourse of Atoms? This also, I believe they will now not much insist upon. Must we then rely on Natural Religion, and the Voice of common Reason for our Guidance to Happiness? This Rule, so far as it can go,

is

is entirely allow'd and improv'd by the Scriptures, and does it self naturally lead us farther to Divine Revelation, as the only Security of not erring in many Cases, especially those of Divine Worship, and Propitiation; and the principal Means of discovering the Certainty of such future Rewards and Punishments, as are in many Cases absolutely necessary to support the Observance of the Laws of Nature. Must we, I say, take our Leaves of the *Jewish* and *Christian* Revelations, which have such Mighty Attestations as coming from God, and trust our selves wholly to our own Human, Frail, and Uncertain Imaginations, Inclinations, and Conjectures in Divine Matters? This seems very hard, and very unreasonable. Let us suppose that the Philosophy and Religion we now are in Possession of, have several difficult Places, and some hitherto unaccountable Phænomena in them: Yet for certain may we justly expect to have a better Philosophy and Religion to betake our selves to, before we leave these; I mean better attested, and freer from Perplexities and Incumbrances, both in Faith and Practice; otherwise we shall act absurdly; and imitate the Folly of him who pulls down his present House, on Account of some apparent Inconveniencies therein, without either the Skill or the Ability to build a better in its Stead. And for my self, I venture to promise all the Scepticks and Unbelievers of our Age, that as soon as ever they will shew me a Scheme of Philosophy and Religion more Rational, and better Attested to than that of the *Christian*, which is the Perfection of the *Jewish* also, I will be their Proselyte: But till then they must excuse me.

These

These, Reader, are some of the Arguments and Motives which induce me really to believe, and conscienciously to endeavour to live up to the *Christian* Institution, and to admit the Holy Scriptures for Divinely Inspired: And I heartily wish they may have the same Influence upon every Reader; that so they may *with the Heart believe unto Righteousness, and with the Mouth make Confession unto Salvation.* Rom. x. 10. One Thing I will add here; that since we cannot act upon Motives and Arguments any farther than we can know them; nor can we go upon Evidence any farther than it is extant in the World for our Examination: And since the Justice of God can only require us to act upon the Motives and Evidence his Providence affords us, and can only call us to an Account for our Behaviour in Agreement with such Motives and Evidence; I venture to say, with great Assurance, that whatever be the Truth of the Things themselves, which we can no otherwise determine about, the Arguments and Evidence now Extant in the World, do, for the main, so greatly and undoubtedly *preponderate* on the Side of the *Jewish* and *Christian* Revelations, that we are bound by all the Rules of Justice, and Reason, and good Sense, to prefer it, to be determin'd by it, and act upon it; and that therefore those who do otherwise may justly be call'd to Account, and punish'd for the contrary Procedure, for their Infidelity and Disobedience; seeing these are chosen in Defiance of plainly greater, and plainly *superior Evidence* for the Divine Authority of the Sacred Writings.

I conclude the whole with Two remarkable Passages, taken from the Two great Apostles St. *Peter* and St. *Paul*, and with the Recommendation

dation of Three Genuine Memorials of the most Primitive Martyrs: and I pray God they may equally affect others, as they always do my self, while I consider them as the Solemn Attestations of such Persons, who most of them certainly knew whether the Religion they Preach'd was Divine or not; and as generally made a little before their Deaths also; when, if ever, Men use to be serious, sincere, and in earnest in such their Declarations.

2 Pet. i. 10—18.

Peter.] *Wherefore the rather, Brethren, give Diligence to make your Calling and Election sure: For if ye do these Things ye shall never fall: For so an Entrance shall be ministred unto you abundantly, into the everlasting Kingdom of our Lord and Saviour Jesus Christ. Wherefore I will not be negligent to put you always in Remembrance of these Things, though ye know them, and be established in the present Truth. Yea, I think it meet, as long as I am in this Tabernacle, to stir you up, by putting you in Remembrance: Knowing that shortly I must put off this my Tabernacle, even as our Lord Jesus Christ hath shewed me. Moreover, I will endeavour that you may be able after my Decease to have these Things always in Remembrance. For we have not followed cunningly devised Fables, when we made known unto you the Power and Coming of our Lord Jesus Christ, but were Eye-Witnesses of his Majesty. For he received from God the Father, Honour and Glory, when there came such a Voice to him from the excellent Glory, This is my beloved Son, in whom I am well pleased. And this Voice which came from Heaven we heard, when we were with him in the Holy Mount.* [See Matt. xvii. 1--13.]

Paul.]

of RELIGION.

Paul.] *I charge thee therefore* [Timothy] *before God, and the Lord Jesus Christ, who shall judge the Quick and the Dead, at his appearing, and his Kingdom: Preach the Word, be instant in Season, out of Season; reprove, rebuke, exhort with all Long-suffering and Doctrine.*

2 Tim. iv. 1, 2.

But watch thou in all things, endure Afflictions, do the Work of an Evangelist, make full Proof of thy Ministry. For I am now ready to be offered, and the time of my departure is at hand. I have fought a good Fight, I have finished my Course, I have kept the Faith. Henceforth there is laid up for me a Crown of Righteousness, which the Lord the righteous Judge shall give me at that Day: And not to me only, but unto all them also that love his appearing.

v. 5—8.

The Genuine Memorials of the Martyrs which I would here recommend, (and they are almost all the Memorials of that Kind, which appear to be very Ancient and certainly Genuine) are, the Epistle of St. *Ignatius*, Bishop of *Antioch*, to the *Romans*, as he was going to Martyrdom: The Epistle of the Church of *Smyrna*, concerning the Martyrdom of St. *Polycarp* their Bishop: And the Epistle of the Churches of *Vienna* and *Lyons*, concerning the Martyrs under the Persecution of *Verus* in *Eusebius*: All which, if they do not in some measure Affect Men, and make them sensible, that the first Christians, even those who certainly knew whether Christianity was true or not, were in earnest, and believed themselves, their Hearts are *as hard as the nether Milstone*, and past all ordinary Ways of Influence and Conviction.

Hist. Eccl. V. 1.

Lyndon in *Rutland*.
Sept. 1. 1716.

WILL. WHISTON.

Sir

Sir *Richard Blackmore*'s HYMN

TO THE

CREATOR.

*H*Ail King Supream! of Pow'r Immense Abyss!
  Father of Light! Exhaustless Source of Bliss!
Thou Uncreated, Self-existent Cause,
Controul'd by no Superior Being's Laws;
E'er Infant Light essay'd to dart the Ray,
Smil'd heav'nly sweet, and try'd to kindle Day;
E'er the wide Fields of Ether were display'd,
Or Silver Stars Cerulean Spheres inlaid;
E'er yet the eldest Child of Time was Born,
Or verdant Pride young Nature did adorn,
Thou Art; and didst Eternity employ
In unmolested Peace, in Plenitude of Joy.

In its Ideal Frame the World design'd
From Ages past lay finish'd in Thy Mind.
Conform to this Divine Imagin'd Plan,
With perfect Art th' amazing Work began.
Thy Glance survey'd the Solitary Plains,
Where shapeless Shade inert and silent Reigns;
Then in the dark and undistinguish'd Space,
Unfruitful, uninclos'd, and wild of Face,
Thy Compass for the World did mark the Place.
Then didst Thou through the Fields of barren Night
Go forth collected in Creating Might.
Where Thou Almighty Vigour didst exert,
Which Emicant did This and That Way dart

Thro'

Thro' the black Bosom of the empty Space:
The Gulphs confess th' Omnipotent Embrace,
And pregnant grown with Elemental Seed
Unfinish'd Orbs, and Worlds in Embryo breed.
From the crude Mass, Omniscient Architect,
Thou for each Part Materials didst select,
And with a Master-hand Thy World erect.
Labour'd by Thee, the Globe's vast lucid Buoys
By Thee uplifted float in liquid Skies.
By Thy cementing Words their Parts cohere,
And roll by Thy Impulsive Nod in Air.
Thou in the Vacant didst the Earth suspend,
Advance the Mountains, and the Vales extend;
People the Plains with Flocks, with Beasts the Wood,
And store with Scaly Colonies the Flood.

Next Man arose at Thy Creating Word,
Of Thy Terrestrial Realms Vicegerent Lord.
His Soul more artful Labour, more refin'd,
And Emulous of bright Seraphic Mind,
Ennobled by Thy Image spotless shone,
Prais'd Thee her Author, and ador'd Thy Throne:
Able to Know, Admire, Enjoy her God,
he did her high Felicity applaud.

Since Thou didst all the spacious Worlds display,
Homage to Thee let all Obedient pay.
Let glitt'ring Stars, that Dance their destin'd Ring
Sublime in Sky, with Vocal Planets Sing
Confed'rate Praise to Thee, O Great Creator King.
Let the thin Districts of the waving Air,
Conveyancers of Sound, Thy Skill declare.
Let Winds, the Breathing Creatures of the Skies,
Call in each vig'rous Gale, that roving flies
By Land or Sea, then one loud Triumph raise,
And all their Blasts employ in Songs of Praise.

While painted Herald-Birds Thy Deeds proclaim,
And on their spreading Wings convey Thy Fame;
Let Eagles, which in Heav'n's Blue Concave soar,
Scornful of Earth superior Seats explore,

And

And rise with Breasts erect against the Sun,
Be Ministers to bear Thy bright Renown,
And carry ardent Praises to Thy Throne.

Ye Fish assume a Voice, with Praises fill
The hollow Rock, and loud reactive Hill.
Let Lions with their Roar their Thanks express,
With Acclamations shake the Wilderness.
Let Thunder-Clouds, that float from Pole to Pole,
With Salvoes loud salute Thee, as they roll.
Ye Monsters of the Sea, ye noisy Waves
Strike with Applause the repercussive Caves.
Let Hail and Rain, let Meteors form'd of Fire,
And lambent Flames, in this blest Work conspire.

Let the High Cedar, and the Mountain Pine
Lowly to Thee, Great King, their Heads incline.
Let ev'ry Spicy Odoriferous Tree
Present its Incense, and its Balm to Thee.

And Thou, Heav'n's Viceroy o'er this World below,
In this blest Task Superior Ardor show:
To view thy Self inflect thy Reason's Ray.
Transported, Nature's Theatre survey.
Then all on Fire the Author's Skill adore,
And in loud Songs extol Creating Pow'r.

Degenerate Minds in mazy Error lost
May combat Heav'n, and impious Triumphs boast;
But while my Veins feel animating Fires,
And vital Air my breathing Breast inspires,
Grateful to Heav'n I'll stretch a pious Wing,
And Sing His Praise, who gave me Pow'r to Sing.
                                    Creation, Lib. VII. in calce.

FINIS.

ERRATA.

Page 51. Line 15. read 130,000. l. 16. r. 364,000. p. 54. l. ult. r. 4248. p. 57. l. 6. r. *as to.* l. 16. r. 231. l. 19. r. *near* 3. l. 22. r. 12. p. 59. Marg. add p. 426, —430. p. 109. l. 28. add in the Marg. De Universo, C. 1. p. 255. l. 22. r. *it is this.*

THE
CAUSE
OF THE
DELUGE
DEMONSTRATED.

BEFORE I proceed to my present *Demonstration of the Cause of the Deluge*, I must premise this, That in my *New Theory of the Earth*, especially as improv'd and corrected in the Second Edition, I have evidently shewn, that in Case a Comet pass'd by, before the Earth, in its annual Course, on the 17th Day of the Second Month, from the Autumnal Equinox, or *Nov.* 28. in the 2349th Year before the Christian Æra, the Phænomena of Nature and History, and particularly the *Mosaick* Account of the Deluge of *Noah*, which are no otherwise to be accounted for, are exactly explain'd; that the Calculations and Proportions, where-ever we can come at them, are on that Hypothesis right, agreeable to one another, to Ancient, especially Sacred History, and to the System of Astronomy; that there are Traces in Ancient Books of a Tradition, that a Comet did appear at the very Beginning of the Deluge; that the very Month and Day mentioned by *Moses* for such its Beginning, is attested to by other Old Records, and, on this Hypothesis, by Astronomical Calculations also: whence I concluded that it was

most

most highly probable, or rather physically demonstrable, that a Comet did pass by at that time, and was, under the Conduct of the Divine Providence, and as his Instrument in punishing a wicked World, the Cause of that Deluge. The only thing wanting was, to demonstrate from the Period of some Comet, and its Situation in the Heavens, Astronomically stated and computed, that such a Comet did actually come by at that very time: which if it could be once shewn, the whole must be own'd as certain, and demonstrated, and all the natural Corollaries therefrom must be allow'd as true, even by the Obstinate and Incredulous. This indeed at first was look'd upon by me as not at all to be expected; since we then barely began to know, or rather strongly to conjecture that Comets did revolve about the Sun in settled Periods, but without being able to determine any one of those Periods. But of late God has so bless'd the Labours of the Learned; and this Part of Astronomy is so much improv'd, especially by the farther Pains and Observations of the great Inventor himself, Sir *Isaac Newton*; whose Name will never be forgotten while Mathematicks and Astronomy are preserv'd among Mankind; and by the laborious Calculations of the acute Dr. *Halley*, on the Principles laid down by the former, that what was a few Years ago almost despair'd of, is now in great Measure discover'd, and we know not only that one Comet has come round three or four times already in later Ages, *viz.* A. D. 1456, 1531, 1607, and 1682, and will no doubt come round again A. D. 1758, as making its Period in about 75 Years; that another has probably come round in the same later Ages twice already, *viz.* A. D. 1532, and 1661; and so is to return A. D. 1789, or 1790, as making its Period in about 129 Years: But, which is the greatest Discovery of all, that the last most remarkable Comet

met, whofe Defcent into our Regions has occafion'd almoft all the modern folid Knowledge we have relating to the whole Cometick Syftem it felf, has alfo feveral times been feen already within the time of certain Records; I mean in the 44th Year before Chrift, and again *A. D.* 531, or 532; and yet again *A. D.* 1106, befides this its laft Appearance *A. D.* 1680, whereby we know that it revolves in about 575 Years. This laft Comet I may well call the *moft remarkable one* that ever appear'd; fince befides the former Confideration, I fhall prefently fhew, that it is no other than that very Comet which came by the Earth at the Beginning of *Noah*'s Deluge, and which was the Caufe of the fame. Now confidering the Premifes, I fhall only have occafion, in order to my prefent Defign, to prove thefe five Things concerning it. (1.) That no other of the known Comets could pafs by the Earth at the Beginning of the Deluge. (2.) That this Comet was of the fame Bignefs with that which pafs'd by at that time. (3.) That its Orbit was then in a due Pofition to pafs by at that Time. (4.) That its defcending Node was then alfo in a due Pofition for the fame Paffage by. (5.) That its Period exactly agrees to the fame time. Or, in fhort, that all the known Circumftances of this Comet do correfpond, and that it actually pafs'd by on or about that very Year, and on or about that very Day of the Year when the Deluge began. All which Things I fhall demonftrate in their Order.

I. None of the other Comets yet known, I mean of the 21 in Dr. *Halley*'s Table and my *Solar Syftem*, could be that which pafs'd by the Earth at the Beginning of the Deluge. This appears by thefe certain Arguments following.

(1.) None of them appear to have been of a due Bignefs: For the Phænomena of the Deluge, as I have

New Theor. 2d Edit. Corol. 2. Lem. 86. & p. 203, 204.

have elsewhere shew'd, require a small one in Comparison of the Earth, whereas the rest of the Comets seem to have been commonly larger than it.

(2.) None of their descending Orbits are duly situate, I mean between 90 and 100 Degrees from *Aries*: which Position is yet absolutely necessary in this Case. For the Precession of the Equinox, which is about 50 Degrees, added to the 46 Degrees that the Earth was distant from *Aries* when the Flood began, must suppose the descending Orbit of the Comet to be now between 90 and 100 Degrees from *Aries*: at which place none of the descending Orbits of the other Comets are now situate; as Dr. *Halley*'s Table, and my Solar System grounded thereon, will readily shew. (3.) None of the other's Nodes are so situate, as is necessary to bring the Comet near enough to our Earth: I mean between 90 or 100 Degrees from *Aries*; and so as to cross the Plane of the Ecliptick very near to the Distance of the Earth from the Sun; as is also plain from the same Table and System. Nay indeed, the wrong Situation of the descending Orbits, noted under the last Head, renders this due Situation of the Nodes plainly impossible. For it being necessary, that the Orbit it self intersect the Ecliptick it self in the 17th Degree of *Taurus*; this cannot possibly be in such a Situation of the Orbit, as that we have already mention'd to belong to all the rest of the known Comets. So that these other Comets were utterly incapable of being instrumental in the Deluge, even tho' their Periods should any of them agree; which yet we know not that any of them do.

II. This Comet was of the same Bigness with that which pass'd by at that time; I mean a very small one, and only 10 times as large as the Moon. This appears by Mr. *Flamsteed*'s Determination of its apparent Diameter, about 20″ when it was nearly

Ubi supra.

ly as far off as the Sun: whereas he fuppofes that of the Moon at the fame Diftance to be about 6". So that if due Allowance be made for that large and denfe Part of the Atmofphere, which hides the Nucleus or Comet it felf from us, fuppofe 7", the Diameter of the folid Body it felf will be only 13" Now the Cube of 13, or 2197, is to the Cube of 6, or 216, as about 10 to 1. Whence it appears, that this Comet is about ten times fo great as the Moon, or ¼ fo great as the Earth, as the real Comet that occafion'd the Deluge ought to be.

III. The defcending Part of the Orbit of this Comet was about the 17th deg. of *Taurus* at the Time of the Deluge, as that of the Comet at the Deluge muft have been. For this defcending Orbit is now in the 2d Degree of *Cancer*; and if we allow 46 Degrees for its apparent Motion fince the Deluge, which is very little different from the real Preceffion of the Equinox, the main, if not only Occafion of it, it will appear to have been in the 17th Degree of *Taurus* at that Time, according to the foregoing Computation.

IV. The Defcending Node of this Comet, which is of the greateft Confideration here, and liable to the greateft Variety of all, does alfo exceeding well agree in the prefent Cafe. For this is now in the 3d Degree of *Cancer*; and if we allow, as before, 46 Degrees for its apparent Motion fince the Deluge, or for the real Preceffion of the Equinox, the main, if not only Caufe of it, it will appear to have been in the 17th of *Taurus* at that Time alfo. Nay, if we allow the leaft Inequality in thefe two Motions, or the leaft Alteration of the Planes either of the Ecliptick or of the Comets Orbit, or of both, as we juftly may, both from the Phyfical Caufes, and Aftronomical Obfervations, we may fuppofe them ftill nearer the Earth's Diftance from the Sun, and fo more exactly fuitable to the Cafe of the Deluge.

V. The

V. The Period of this Comet most exactly agrees to the same Time, I mean to 7 Revolutions in 4028 Years, the Interval from the Deluge till its last Appearance 1680. For, as Sir *Isaac Newton* first observ'd, from its Elliptick Curvature before it disappear'd, that its Period must be in general above 500 Years; so has He and Dr. *Halley* since observ'd, that the same Comet has been seen four times, *viz.* the 44th Year before Christ. *A. D.* 531 or 532, *A. D.* 1106, and *A. D.* 1680, and that by consequence it makes a Revolution in about 575 Years. Now if we make a very small Allowance for the old Periods before Christ, and suppose that, one with another, it has revolv'd in $575\frac{1}{2}$ Years, we shall find that 7 such Periods amount to 4028 Years, exactly, according to that Number since the Deluge. This is so remarkable an Observation, and so surprizing, that it will deserve a particular Demonstration from the original Authors themselves. To begin then with the first of the Appearances recorded in later History, I mean that in the 44th Year before Christ, the Year that *Julius Cæsar* was slain, we have no fewer nor lesser Persons than *Seneca, Suetonius, Plutarch* and *Pliny* to attest it; and the last, as bringing *Augustus*'s own Words for his Voucher. Take the Account in those Words, as being the most authentick and remarkable. ' On those very Days, says *Augustus,* when
' I was exhibiting some Games to the People, [be-
' gun about *Sept.* 26.] a Comet appear'd for 7 Days,
' and was seen in the *Northern* Part of Heaven. It
' rose about the 11th Hour of the Day: It was a
' remarkable one, and visible all over the World.
' The common People believ'd, that it signify'd the
' Reception of the Soul of *Cæsar* into the Number
' of the immortal Gods. On which Account the
' Image of this Star was added to that Statue repre-
' senting *Cæsar*'s Head, which we a while after con-
' secra-

Princip.
2d Edit.
p. 465.

Sen. Nat.
Quæst. L.
VII. C.17.
Se on. in
Jul C.88.
Plut. in Cæ-
sar. Plin.
Hist. Nat.
L.II.C.24.
Grut. ap.
Usser.
Annal.

'secrated in the Forum". Accordingly it is known that some of *Cæsar*'s Coins have a Star upon them, for a Memorial of this Comet; and observable that *Virgil* hints at the same also, *Patrium aperitur ver-* *Æneid.* *tice sidus*. *Plutarch*'s, *Seneca*'s and *Suetonius*'s Words are VIII. almost the very same that are included in the Passage from *Augustus*, and so need not be distinctly set down. Only the Time of its Rising is by *Suetonius* set down *about the* 11*th Hour*, without the Words *of the Day*, which the other two have; and its *Northern* Position is only mentioned by *Augustus* himself. Now if we interpret the 11th Hour, or 11th Hour of the Day, to be either 11 a-clock before Noon, or an Hour before Sun-set, this will render the whole almost incredible: it being next to impossible, that this Comet should be seen in the Day-time. But the *Romans* then accounting Midnight the Beginning of their Day, as is well known by Chronologers, we may reckon this 11th Hour to be 11 at Night, and all will agree to the Comet before us; and it will shew, that as it had been conceal'd by cloudy Weather for some time, so it now appear'd ascending from the Sun, with its long and splendid Tail for a Week, before the like cloudy Weather, or the Comet's too great Remoteness rendred it no longer observable. Accordingly the *Northern* Position of this Comet, noted here by *Augustus*, secures us still farther, that it must have been the same with that *A.D.* 1680, which is ever in the same Position, at the same Place of its Orbit: to say nothing of its remarkable Brightness, which I take to belong to its Tail, and which rendred it so very remarkable then in the World: In which Point it as well or better agrees with this, than with any other in the whole Cometary System. So that on all these Accounts, the Comet seen then by the *Romans*, and that seen *A.D.* 1680, must have been one and the same Comet. The next Period

when

when this Comet might be seen again, according to the foregoing Time of its Revolution, was *A.D.* 531, or 532. When yet we hear nothing of it in *Hevelius*'s History of Comets. But then we have it in *Lubienietz*'s more exact Catalogue, out of *Zonaras*, the Original Historian, whose Words are these, Annal. L. xiv. p. 61. 'In the 5th Year of the Emperor *Justinian* [*A. D.* 531, or 532.] a Comet appear'd, of that Sort which is called *Lampadias*. 'It sent its bright Tail upward, and continued to 'shine 20 Days. Which Words exactly agree to this Comet. The next Period when it was to be expected, was *A.D.* 1106. at which Time the Historians are full of their Accounts of it. Take those Accounts in their own Words, as they stand in *Hevelius* and *Lubienietz*, who have given us a most compleat Collection of them in their Histories of Comets.

Lavath. ex Urspurg. p. 148.
Lavath. ex Chron. Norimb. vel aliunde.
Calvis. ex Tyr.
Myzald. p. 140.
Sigebert.
Func.

A. D. 1106. We saw a Comet of wonderful Brightness, from the first Week in *Lent*, until the Passion of our Lord. An extraordinary Star was seen to shine this Year on *Friday* in the Evening, *Southward* and *Westward*, and appeared bright for 25 Days together, and always at the same Hour.

A. D. 1106. in the Month of *February*, two Days after the New Moon, a great Comet appear'd *South-Westward*. *A. D.* 1106. a Comet appear'd like a Fire, almost all the Month of *February*.

A very great Comet was seen in the Time of *Lent*. *Prætorius* adds, that the Emperor *Henry* IV. died the same Year; which *Calvisius* also agrees to.

Append. Marian. Scoti.

A. D. 1106. a Star, which we call a Comet, appear'd.

Exstorm. ex Chron. Saxon.
Hist. Eccl. ex Simeone Dunelmen.

A. D. 1166. a dreadful Comet appear'd, from the first Week in *Lent*, till the Vigil of *Palm-Sunday*. The same Year the Emperor *Henry* IV. died.

On the Year of our Lord 1106, the 14th of the Calends of *March*, [*Feb.* 16.] a certain strange Star was

Demonstrated.

was discovered, and was seen to shine between the *South* and *West* for 25 Days, after the same manner, and at the same Hour. It seemed to be small and obscure; but that Light which went out from it was exceeding bright, and a Splendor, like a great Beam, proceeded from the *East* and *North*, and shot it self upon the same Star.

In these Testimonies, we may see that all the Circumstances of this Comet agree to that of *A. D.* 1680. I mean the Smallness and Obscurity of its Nucleus, the Brightness and Remarkableness of its Tail, its Position *South-West*, and the Direction of its Tail *North-East*. So that there is no Reason to doubt, but it was the very same. Only we must here note, that these two Periods were, one with another, three Quarters of a Year shorter than the last Period. For from *September*, in the 44th Year before Christ, till *February* or *March A. D.* 1106. are but $1148\frac{1}{2}$ Years, or two Periods of $574\frac{1}{4}$ a piece, one with another: whereas from the same *February* or *March* A. D. 1106. till *February* or *March* $168\frac{0}{1}$, when this Comet was about the same Position again, there are just 575 Years. It is rather a Wonder, that the three last Periods of our Famous Comet are so very nearly equal, than that there is this small Inequality among them. Nor is it, by the way, any Wonder therefore, that the four first Periods after the Deluge are to be suppos'd one with another rather above 576 Years to agree exactly to that Time. 'Tis rather a Question whether the rest of the Comets Periods will prove any of them near so equal in Proportion, as even that Allowance makes these to be. Accordingly, Sir *Isaac Newton* and Dr. *Halley* rightly observe, that these Cometary Orbits are the most easily and sensibly disturb'd by the occasional Nearness of their Comets to other Bodies of all others; and so con-

Princip. p. 480. *Prælect. Physico. Math.* p. 358, 359.

considerable Inequalities are to be expected among them.

Note, (1.) That it is highly remarkable, that this is the only Comet yet known, whose Node renders it capable of approaching very near the Body of the Earth; and that the same Node is still so little remote from the Earth's Orbit, as Dr. *Halley* well observes, that it brought this Comet about as near to the same as the Moon this very last time. Hear his remarkable Words, and consider the Consequence of them in this Matter. ' No Comet, *Synops. Comet. in calce.* ' says he, has hitherto threatned the Earth with a ' nearer Appulse than that of 1680. For by Cal-' culation, I find that *November* 11th 1^{h} $6'$ after ' Noon, that Comet was not above a Semidiame-' ter of the Sun, (which I take to be equal to the ' Distance of the Moon) to the *Northwards* of the ' Way of the Earth. At which time, had the ' Earth been there, the Comet would, I think, ' have had a Parallax equal to that of the Moon'. Nor can I pass over his following Words without setting them down, they are so apposite to my present Purpose. ' The former Observations, says ' he, are to be suppos'd as spoken to Astronomers. ' But what might be the Consequences of so near ' an Appulse, or of a Contact, or lastly of a Col-' lision of these celestial Bodies, (which are none ' of them impossible) I leave to be discuss'd by ' the Philosophers.

(2.) Since this Comet's Period is 575 Years, its midde Distance must be about 5,600,000,000 Miles from the Sun; its longer Axis and greatest Distance twice so long, or nearly 11,200,000,000 Miles; its Aphelion Distance about 14 times as great as the Distance of *Saturn*; its greatest Distance to its least, as above 20,000 to 1; and so its greatest Light and Heat to its least, as above 400,000,000 to 1.

(3.) Since

Demonstrated. 13

(3.) Since 575 Years appears to be the Period of the Comet that caus'd the Deluge, what a Learned Friend of mine, who was the Occasion of my Examination of this Matter, suggests, will deserve to be considered, *viz.* Whether the Story of the Phœnix, that celebrated Emblem of the Resurrection in Christian Antiquity; [that it returns once after 5 Centuries, and goes to the Altar and City of the Sun, and is there burnt; and another arises out of its Ashes, and carries away the Remains of the former, *&c.*] be not an Allegorical Representation of this Comet; [which returns once after 5 Centuries, and goes down to the Sun, and is there vehemently heated, and its outward Regions dissolv'd; yet that it flies off again, and carries away what remains after that terrible burning, *&c.*] and whether the Conflagration and Renovation of things, which some such Comet in its Ascent from the Sun may bring upon the Earth, be not hereby prefigur'd. I will not here be positive; but I own that I don't know of any Solution of this famous Piece of *Egyptian* Mythology and Hieroglyphicks, as this seems to be, that can be compared with it.

Note, (4.) That none of those Comets whose Orbits are yet known, can come near enough to our Earth in their Ascent from the Sun to cause the Conflagration. This is evident to those who consider Dr. *Halley*'s Table, or my Solar System built upon it; since none of them move in or very near the Plane of the Ecliptick; and those four which have their Nodes nearest the Earth's Orbit, and so might approach nearest to the Earth, are either such as have these Nodes so near only in their Descent to the Sun; as that in 1472, and that in 1618, and that in 1680; or go not any time much nearer to the Sun than the Earth it self, as that in 1684. and so are on all Accounts utterly incapable of affording

fording. Heat enough for such a Conflagration.

Note (5.) That therefore the Period or Time for that Conflagration, upon the Supposition that it is to be caused by a Comet, cannot now be discover'd by any natural Means; but must still remain, as formerly, only knowable from Divine Revelation.

New Theory, p. 450, -453.

Note (6.) That hence those remarkable Corollaries, drawn from the accurate Solution of such Difficulties now, as formerly were plainly insoluble; I mean, the great Regard due to the Ancientest Sacred and Prophane Records, and to the inspired Method whence they must have been deriv'd; the Imperfection of Human Knowledge; the Folly of rejecting Revealed Truths, out of regard to uncertain Human Reasonnings; the Wisdom of adhering to the most obvious Sense of Scripture; the Reasonableness of believing Scripture-Accounts and Scripture-Mysteries, tho' not fully comprehended by us; the Justness of expecting Satisfaction in moral Difficulties in due time from the like Satisfaction afforded already in those that are Philosophical, and the like, do all receive a new and surprizing Confirmation; and will therefore deserve a new and serious Consideration.

N. B. *Dr. Halley having himself given an Account of this Comet lately in Dr. Gregory's English Astronomy, P. 901, 902, 903, I here present it to the Reader verbatim, that he may compare the two Accounts together, for his more entire Satisfaction.*

" But as far as Probability from the Equality
" of Periods, and similar Appearance of Comets,
" may be urged as an Argument, the late won-
" drous Comet of 168⁹⁄₈, seems to have been the
" same, which was seen in the Time of our King
" *Henry* I. *Anno* 1106, which began to appear in
" the

" the *West* about the middle of *February*, and con-
" tinued for many Days after, with such a Tail as
" was seen in that of 1680/1. And again in the Con-
" sulate of *Lampadius* and *Orestes*, about the Year
" of Christ 531, such another Comet appeared
" in the *West*, of which *Malela*, perhaps an Eye-
" witness, relates that it was μέγας κ) φοβερὸς, *a great
" and fearful Star*; that it appeared in the *West*, and
" emitted upwards from it a long white Beam;
" and was seen for 20 Days. It were to be wish'd
" the Historian had told us what Time of the Year
" it was seen; but 'tis however plain, that the
" Interval between this and that of 1106, is near-
" ly equal to that between 1106 and 1680/1, *viz.*
" about 575 Years. And if we reckon backward
" such another Period, we shall come to the
" 44th Year before *Christ*, in which *Julius Cæsar*
" was murder'd, and in which there appear'd a
" very remarkable Comet, mentioned by almost
" all the Historians of those Times, and by *Pliny*
" in his Natural History, *lib.* 11. *c.* 24. who recites
" the Words of *Augustus Cæsar* on this Occasion,
" which lead us to the very Time of its Appear-
" ance, and its Situation in the Heavens. These
" Words being very much to our purpose, it may
" not be amiss to recite them. *In ipsis Ludorum me-
" orum diebus, sydus crinitum per septem dies, in regione
" Cæli quæ sub Septentrionibus est conspectum. Id orieba-
" tur circa undecimam horam diei, clarumq; & omnibus
" terris conspicuum fuit.* Now these *Ludi* were de-
" dicated *Veneri genetrici*, (for from *Venus* the *Cæ-
" sars* would be thought to be descended,) and be-
" gan with the Birth-day of *Augustus*, viz. *Sept.* 23.
" (as may be collected from a Fragment of an
" Old *Roman* Calendar extant in *Gruter*, pag. 135.)
" and continued for 7 Days, during which the Co-
" met appeared. Nor are we to suppose that it was
" seen

" seen only those 7 Days, but possibly both before
" and after. Nor are we to interpret the Words
" *sub Septentrionibus*, as if the Comet had appear'd
" in the *North*, but that it was seen under the *Sep-
" tem triones*, or brighter Stars of *Ursa major*. And
" as to its rising *Hora undecima diei*, it can no ways
" be understood, unless the word *diei* be left out,
" as it is by *Suetonius*; for it must have been very
" far from the Sun, either to rise at Five in the
" Afternoon, or at Eleven at Night; in which
" Cases it must have appeared for a long time, and
" its Tail have been so little remarkable, that
" it could by no means be call'd, *Clarum & om-
" nibus Terris conspicuum Sydus*. But supposing this
" Comet to have traced the same Path with that
" of the Year 1680, the ascending part of the Orb
" will exactly represent all that *Augustus* hath said
" concerning it; and is yet an additional Argu-
" ment to that drawn from the Equality of the
" Period. Thus 'tis not improbable but this Co-
" met may have four times visited us at Intervals
" of about 575 Years: Whence the Transverse
" Diameter of its Elliptic Orb will be found
" $\sqrt[3]{575\times575}$ times greater than the annual Orb;
" or 138 times greater than the mean Distance of
" the Sun; which Distance, tho' immensely
" great, bears no Proportion to that of the Fixed
" Stars.

F I N I S.

A Compleat Catalogue of Mr. WHISTON's *Writings, according to the Order of Time when they were Publish'd.*

ENGLISH.

(1.) A New Theory of the Earth, from the Creation to the Consummation of all Things. 2d Edition, with great Corrections and Improvements. 8*vo*. Price bound 6 *s*.

(2.) The Chronology of the Old Testament, and the Harmony of the Four Evangelists. 4*to*. 8 *s*.

(3.) An Essay on the Revelation of St. *John*; with Two Dissertations at the End. 4*to*. 7 *s*.

(4.) The Fulfilling of Scripture Prophecies, in Eight Sermons at Mr. *Boyle's* Lecture; with a Supplement and a Postscript. 8*vo*. 3 *s*. 6 *d*.

(5.) A Memorial for setting up Charity-Schools in *England* and *Wales*. Half a Sheet. Given *Gratis*.

(6.) Sermons and Essays on several Subjects; with *Novatian De Trinitate*. 8*vo*. 4 *s*. 6 *d*.

(7.) Collection of Small Tracts against Dr. *Alix*, Dr. *Grabe*, Dr. *Smallbroke*, &c. 8*vo*. 2 *s*. 6 *d*.

(8.) Primitive Christianity Reviv'd: In Five Volumes. (1,) An Historical Preface. A Dissertation on the Epistles of *Ignatius*, with the Epistles themselves, *Greek* and *English*, and *Eunomius's* Apologetick. (2.) The Constitutions of the Holy Apostles, *Greek* and *English*. (3.) A Vindication of those Constitutions. (4.) An Account of the Primitive Faith; with the Fourth

B Book

A Catalogue of

Book of *Esdras*, from the *Latin* and *Arabick*. (5.) The Recognitions of *Clement*. To all which may be added, a Collection of small Tracts relating to them, but not therein contain'd. 8*vo*. 1 *l*. 13 *s*.

(9.) The Suppofal; or a New Scheme of Government. Half a Sheet. Given *Gratis*.

(10.) *Athanafius* convicted of Forgery. 8*vo*. 3 *d*. *Note*, that this is more compleat, with its Vindication, at the End of the *Argument*, afterward published.

(11.) Primitive Infant-Baptism reviv'd. 8*vo*. 6 *d*.

(12.) Proposals for Erecting Societies for promoting Primitive Christianity. Half a Sheet. Given *Gratis*.

(13.) Primitive Christianity reviv'd: The Four Volumes in one; all *Englifh*. 8*vo*. 7 *s*. 6 *d*.

(14.) Dr. *Mather's* Old Paths reviv'd; with a New Preface. 12*o*. 3 *d*.

(15.) A Scheme of the Solar Syftem, with the Orbits of the 21 Comets. In a large Sheet, engrav'd on Copper, by Mr. *Senex*. 2 *s*. 6 *d*.

(16.) Reflexions on a Difcourse of Free-Thinking. 2d Edition. 8*vo*. 8 *d*.

(17.) Three Effays. (1.) The Council of *Nice* vindicated from the *Athanafian* Herefy. (2.) A Collection of Ancient Monuments thereto relating. (3.) The Liturgy of the Church of *England* reduc'd nearer to the Primitive S andard. 8*vo*. 4 *s*. 6 *d*.

(18.) An Epitome of the Effay on the Revelation: In a Copper Plate explain'd. 6 *d*.

(19.) The Chriftian's Rule of Faith: Or a Table of the moft Ancient Creeds. Engraved in Copper. 1 *s*.

(20.) An Argument concerning the Diffenters Baptifm, and other Miniftrations: With Two Appendices. 8*vo*. 8 *d*.

(21.) Letters to Dr. *Sacheverel*, and Mr. *Lydal* his Affiftant. Given *Gratis*.

(22.) The Caufe of the Deluge demonftrated. An Appendix to the New Theory. 3d Edit. 8*vo*. 3 *d*.

(23.) A

Mr. Whiston's *Writings*.

(23.) A Course of Mechanical, Optical, Hydrostatical, and Pneumatical Experiments, perform'd by Mr. *Whiston*, and Mr. *Hauksbee*. 4*to*. 5 *s*.

(24.) A New Method for Discovering the Longitude. The 2d Edition, with great Improvements. By Mr. *Whiston* and Mr. *Ditton*. 8*vo*. 1 *s*.

(25.) His Defence, prepared for the Court of Delegates. With his Reasons against that Procedure. 8*vo*. 3*s*.

(26.) The *Copernicus*: Describing an Astronomical Instrument so called. 12*o*. 1 *s*.

(27.) A Vindication of the *Sibylline* Oracles. 8*vo*. 2 *s*. 6 *d*.

(28.) An Account of the last, and of the next great Eclipse of the Sun; engraved in Copper. By it self 2 *s*. 6 *d*. But Roll'd, with Mr. *Whiston's* Second, and both Dr. *Halley's* Schemes. 7 *s*.

(29.) St. *Clement's* and St. *Irenæus's* Vindication of the Apostolical Constitutions; with a large Supplement to the 2d Edition. 8*vo*. 1 *s*.

(30.) An Account of the surprizing Meteor, seen *March* 6. 171$\frac{5}{6}$. 8*vo*. 1 *s*.

(31.) An Address to the Princes of *Europe*, for the Admission, or at least the Open Toleration of the Christian Religion in their Dominions. 8*vo*. 1 *s*.

(32.) Astronomical Principles of Religion, Natural and Reveal'd. 8*vo*. 5 *s*.

Now in the Press,

(33.) A Commentary on the Three Catholick Epistles of St. *John*. 8*vo*. 1 *s*.

Preparing for the Press,

(34.) Scripture Politicks: Or an Impartial Account of the Origin and Measures of Government, Ecclesiastical and Civil, from the Books of the Old and New Testament. To be Dedicated to the Right Reverend the Lord Bishop of *Bangor*.

A Catalogue of, &c.

Published by Mr. Whiston,

(1.) Mr. *Chub*'s Supremacy of the Father; in Eight Arguments. 8vo. 1s.

N. B. The same Author has lately Published Two Enquiries: The One concerning Property, or of Liberty of Conscience: The Other concerning Sin, or of Original Sin. 8vo. 1s. Sold by Mr. *Roberts* in *Warwick-Lane*.

Now in the Press,

The Primitive Catechism: Useful for Charity-Schools. By a Presbyter of the Church of *England*. 8vo. 1s.

LATIN.

(35.) Prælectiones Astronomicæ, Cantabrigiæ in Scholis Publicis habitæ. 8vo. 5s. 6d.

(36.) Euclidis Elementa, juxta editionem Cl. Tacquetti: cum additamentis. 8vo. 4s. 6d.

(37.) Prælectiones Physico-mathematicæ, sive Philosophia Newtoni Mathematica. 8vo. 4s. 6d.

Publish'd by him,

V. C. Algebræ Elementa. 8vo. 4s. 6d.

The 35th, 36th and 37th, are also publish'd in English, *under the Author's Review.*

N. B. These Books are Sold by the Author Himself in *Cross-street, Hatton-Garden*; or by Mr. *Tooke* near *Temple Bar*; or by Mr. *Clarke* in the *Poultrey*; or by Mr. *Senex* at the *Globe*, near *Salisbury-Court, Fleetstreet*; and Mr. *Taylor* at the *Ship* in *Pater-Noster-Row*; or by Mr. *James Roberts*, the Publisher, in *Warwick-Lane, London*: or by Mr. *Crownfield* at the University-Press, *Cambridge*: For some of whom they were all Printed. His Astronomical Instrument, called the *Copernicus*, is Sold by Himself, and Mr. *Senex*; as also by Mr. *Hudson*, at the Cabinet in *Frith-street*, near *Soho*-Square. Price 6 Guineas.

March 25. 1717.   *W. W.*